TIME HORIZONS AND TECHNOLOGY INVESTMENTS

Committee on Time Horizons and Technology Investments

NATIONAL ACADEMY OF ENGINEERING

NATIONAL ACADEMY PRESS
Washington, D.C 1992

NATIONAL ACADEMY PRESS • 2101 Constitution Ave., NW • Washington, DC 20418

NOTICE: The National Academy of Engineering was established in 1964, under the charter of the National Academy of Sciences, as a parallel organization of outstanding engineers. It is autonomous in its administration and in the selection of its members, sharing with the National Academy of Sciences the responsibility for advising the federal government. The National Academy of Engineering also sponsors engineering programs aimed at meeting national needs, encourages education and research, and recognizes the superior achievement of engineers. Dr. Robert M. White is president of the National Academy of Engineering.

This publication has been reviewed by a group other than the authors according to procedures approved by a National Academy of Engineering report review process.

Funding for this effort was provided by the the National Academy of Engineering Fund.

Library of Congress Catalog Card Number 92-80242
International Standard Book Number 0-309-04647-5

Additional copies of this publication are available from:
National Academy Press
2101 Constitution Ave, N.W.
Washington, D.C. 20418

S489

Printed in the United States of America

COMMITTEE ON TIME HORIZONS AND TECHNOLOGY INVESTMENTS

DONALD N. FREY, *Chairman,* Professor of Industrial Engineering and Management Sciences, Northwestern University
ROBERT C. FORNEY, Retired Executive Vice President, E. I. du Pont de Nemours & Company
MARTIN GOLAND, President, Southwest Research Institute
GEORGE N. HATSOPOULOS, Chairman and President, Thermo Electron Corporation
TREVOR O. JONES, Chairman of the Board, Libbey-Owens-Ford Company
HENRY KRESSEL, Managing Director, Warburg, Pincus & Company
JOHN R. MOORE, Retired Vice President and General Manager, Electro-Mechanical Division, Northrop Corporation
JOHN W. PODUSKA, SR., President and CEO, Stardent Computers Inc.
JAMES BRIAN QUINN, William and Josephine Buchanan Professor of Management, Amos Tuck School of Business, Dartmouth College
SHELDON WEINIG, Chairman and CEO, Materials Research Corporation

Staff

BRUCE R. GUILE, Director, NAE Program Office
KATHRYN J. JACKSON, NAE Fellow

Preface

It is often asserted that the short time horizons of U.S. firms are contributing to a decline in the technological strength and competitive performance of U.S. companies in technology-intensive industries. Although this assertion is now widely accepted among those concerned with U.S. competitive performance, relatively little attention has been focused on examining, in detail, the time horizons that organizations use or the forces that create inappropriate, short time horizons.

To address these issues, the National Academy of Engineering established a Committee on Time Horizons and Technology Investments under the chairmanship of Donald Frey, formerly chairman and CEO of Bell & Howell Corporation and now professor of industrial engineering and management sciences, Northwestern University. The study committee was carefully balanced to ensure that the members of the committee, individually and collectively, had deep expertise in matters of corporate governance, investment decision making, technology commercialization, manufacturing, management of research and development, and general business management.

A study was launched to gather evidence and sort out claims about the time horizons of U.S. businesses and the impact of time horizons on the willingness of companies to invest in research, development, or the deployment of new technology. The particular concern was that the time horizons of U.S. businesses were shortening and that shortened time horizons would create a reluctance to invest in technology-related activities with long-term payoff.

The study had a special focus on the interaction of corporate governance and managerial decision making with the overall financial and economic environment. Although focusing on the technological aspects of businesses, the study nonetheless needed to engage issues such as the cost of capital, financial market structures, the impacts of mergers and acquisitions, changing company capital structures, and increasing debt levels. The committee's report makes an important contribution to a complex and often confused aspect of debates over the causes of, and solutions to, U.S. competitiveness problems.

On behalf of the Academy of Engineering I would like to thank the chairman and the members of the committee (p. iii) for their insights and efforts on this project. Over the course of two years, several committee meetings and workshops, and innumerable FAXs, phone calls, and draft versions of the report, they remained actively engaged and unfailingly constructive. I would also like to thank several members of the NAE Program Office. NAE Fellow Kathryn J. Jackson served as study director from the start of the study through the spring of 1991, when she left at the end of her fellowship. Annemarie Terraciano and Margery Harris provided administrative and logistical support for the project. Bruce R. Guile, director of the NAE Program Office, added this study to his workload in mid-1991 and saw it through to completion. They all deserve special thanks.

Robert M. White
President
National Academy of Engineering

Contents

TIME HORIZONS AND TECHNOLOGY INVESTMENTS

Executive Summary

In early 1990 the National Academy of Engineering formed a Committee on Time Horizons and Technology Investments (p. iii) to explore the impact of time horizons on the development and deployment of both product and process technologies by U.S. corporations. The committee solicited, received, and discussed both written and oral background and insights, but since the study, by design, explored an undeveloped and poorly understood issue, the findings and recommendations in this report draw relatively more on the experience and insights of the committee members and relatively less on earlier empirical or theoretical work.

The committee concluded that there is clear justification for concern over U.S. corporate time horizons. There are significant numbers of industries, or segments of industry, in which short-horizon behavior seems to be both the norm and a considerable source of competitive disadvantage. In addition, there is macroeconomic evidence—low relative rates of investment in long-lived assets and in R&D—that appears to indicate a broad-based tendency toward short-term planning and performance criteria on the part of U.S. industry.

However, the committee found that the common presentation of time horizons in U.S. companies—that U.S. executives are generally near-term oriented and pursue only short-term goals—is too simplistic. *Time horizons for technology investments should, and do, vary widely by industry, product, and business activity.* It is also clear that short time horizons are not a universal problem for U.S. companies; there are a number of successful U.S. companies operating in industries that require relatively long time horizons for investments.

Near-term orientation in a company can be characterized as a preference for a portfolio of investments that are likely to yield returns in the near future. In many cases, such preferences are rational reflections of technological and marketplace uncertainty and the investment risk they create. These are natural countervailing forces to longer-term planning and investing. This link between risk and time horizons is also quite explicit in the role that capital costs and investment hurdle rates (the discount rates used in company decision making) play in investment decision making in companies; the more risky the project or venture, the more likely it is that both financial markets and internal management decision-making processes will require a higher expected return.

The relationship between risk and investment time horizons is particularly important with regard to investments in the development and deployment of new product or process technologies. Investments in technology-dependent ventures may, in early years, create largely intangible assets, investments may be illiquid for long periods of time as projects can be slow to mature, and they are exposed to both normal business risk and technology-related uncertainty. As a result, technology investments often carry a substantial (formal or informal) risk premium. Although some of the risk is irreducible, a substantial portion reflects the capability of a company in bringing a competitive new product to market or in introducing a substantial process innovation. This implies that adoption of short time horizons in technology-dependent investments is a result of a company's inability to manage technology effectively. *Companies with deep and genuine competence in commercial application of technology will have a distinct advantage in adopting longer time horizons for technology investments because they are able to reduce the risk of those investments.*

Important aspects of any company's options, practices, and time horizons are also created by (1) the specific competitive status of a company, marketplace and technological uncertainty, and the abilities of a company's board of directors and executive managers to deal with uncertainty; (2) the expectations of investors (the cost and patience of capital) and the way those expectations interact with the financial structure and investment practices of a company; and (3) the design and implementation of government policy. The diversity of influences on corporate time horizons strongly imply that no single actor can unilaterally lengthen investment time horizons. The federal government, boards of directors, and company management *all* need to act if U.S. technology investment time horizons are to be lengthened.

Boards of directors and the top management they select are uniquely responsible for a company's future. Thus, if a public company's performance is weak because of shortsighted investment behavior, it is ultimately a failure of its board of directors. The board of directors significantly influences the time horizons of a company through the selection and development of senior management. To be effective at governing a corporation, a board of directors should be attentive to the importance of balance within the senior management team with regard to (1) the age distribution of senior management; (2) the degree to which there are high-quality, identifiable champions for the initiatives that should mature into a company's core businesses in the next decade; and (3) the balance between members of the senior management team focusing on near-term problems and those focusing on the long-term future of the company. Boards of directors should link compensation packages for their senior executives to their performance in developing and implementing plans for the long-term performance of the company. Also, because of the special characteristics of ventures or plans that depend heavily on the use of technology, it is crucial that boards of directors understand commercial technological innovation.

Since the actions of boards of directors are crucial to the time horizons of a company, so are the methods by which directors are selected, compensated, and removed. First, selection of board members should not be an exclusive prerogative or responsibility of the chief executive officer (CEO). Second, corporate governance might be improved by increasing the financial stake that outside directors have in the corporation by requiring that they own shares at least equal in value to a specified multiple of their annual fees as directors. This measure is intended to link board member compensation as directly as possible to long-term stock performance. It is important to note that directors' compensation schemes are not a cure-all for the many perceived ills of boards of directors of public companies.

> *The committee recommends that corporate boards have nominating committees operating independently of the CEO in choosing new board members and that these nominating committees, in technology-driven companies, give more weight to technological skill as well as business experience in selecting new board members.*

> *The committee recommends that corporations move to increase the financial stake that their directors have in the corporation and that a significant part of directors' compensation be paid in stock or stock options.*

Senior management plays a very important role with regard to time horizons in at least three ways: (1) constancy of purpose coupled with flexibility in the development and execution of corporate strategy; (2) design and implementation of career development systems and compensation schemes that promote attention to longer-term corporate goals; and (3) design and choice of decision-making methods and measurement tools that suit the demands and uncertainties of technology-dependent investments.

The committee recommends that a greater portion of the compensation of managers who are in a position to influence the long-range technological performance of a corporation be granted in stock or stock options. The options should not be exercisable for several years—perhaps five years—and should last for a number of years—perhaps ten years.

The committee recommends that bonuses paid to managers with scope and authority over long-term performance be based not just on the previous year's performance, but on multiple years' accomplishments.

The committee recommends that companies actively reconsider the way they use investment decision-making tools such as discounted cash flow analysis, especially with regard to decisions involving new or continuing investments in technology development and deployment. Faulty or unrecognized implicit assumptions, lack of attention to strategic considerations, and poor handling of technological or market uncertainty in the use these tools can critically damage a company's decision making about technology investments.

The emergence of large institutions as important factors in corporate ownership is an important change affecting the pace and character of restructuring and redirecting U.S. industrial enterprises, with possible effects on corporate time horizons. However, even in the context of this trend in capital markets, corporate senior managers continue to have some influence over a company's cost of capital, and thereby exercise control over one determinant of investment time horizons. Managements and boards of directors can (1) affect how markets perceive the firm's potential as an investment opportunity by a variety of actions, including those that establish long-term relationships with key participants, (2) control the capital structure and have some influence on the ownership structure of the company, and (3) take advantage of opportunities for project or venture risk sharing to reduce capital costs.

The committee recommends that managements and boards of directors of companies dependent on long-horizon technological developments (a) implement investor-relations strategies that aggressively and clearly communicate the technological prospects of a company; (b) work to develop long-term relationships with lenders and equity investors; and (c) aggressively pursue joint ventures or other arrangements to reduce the risk of specific technological ventures.

With regard to international differences in the market cost of capital (defined here as financial investors' required expected return), national differences in rates for debt are not likely to exist except for relatively short-lived fluctuations arising from national economic or monetary policies. However, national differences in the market cost of equity are likely to persist at some level, and U.S. companies should prepare themselves to operate with some disadvantage in this area.

The committee recommends that the federal government move to allow longer investment time horizons for U.S. corporations through tax policy changes designed to reduce the pretax cost of equity capital.

Other government polices and investments also have a pervasive, important, and often positive influence on the business environment and economic development of the United States. Of these influences, this report deals only with the impact of government investments and regulatory policies on investment time horizons.

Regulations and legal procedures can either increase or decrease the risk faced by private investment. As such, some regulations lengthen corporate time horizons, while other regulations, or legal constraints that introduce substantial unpredictability, can cause firms to shorten their time horizons. The importance of government policies with regard to the regulation and creation of markets needs to be acknowledged, and capability in the use of such policies to support long-term investment should be strengthened.

The committee recommends that the federal government invest in improving the efficiency and timeliness of its regulatory, patent, and licensing procedures.

With regard to government investments, the government creates complementary assets—publicly provided infrastructures or services that permit, support, or work in conjunction with private investments in physical or human capital or R&D. Such assets can reduce the risk of related private investments and allow private companies to adopt longer time horizons for their investment decisions. Publicly supported research and development and public infrastructure are two primary examples.

The committee recommends that the budgetary process for the federal government include more explicit consideration of the degree to which federal expenditures support the creation of long-lived physical and human capital or a knowledge base. Preference should be given to those expenditures that will generate returns for long periods of time and contribute to lengthening the time horizons of private-sector investments in the development and deployment of technology.

1

The Issue and the Approach

All too often American management is under pressure to improve the bottom line in the next quarter, without regard to how their actions will affect business in the future. This problem is not simply a result of myopic management; it is systemic. The cost of capital, budget and trade deficits, the tax system and the pressure of financial markets all contribute to the problem.

—*Council on Competitiveness,* Picking Up the Pace: The Commercial Challenge to American Innovation *(Washington, D.C., 1988), p. 11.*

It is frequently argued that U.S. technology-intensive corporations have shorter time horizons for planning and investment than do their principal Japanese and German competitors. As such, the near-term orientation of U.S. companies is often cited as the headwaters of a cascading sequence of events that threaten U.S. economic welfare:

- Companies with time horizons that are too short invest too little in the development and deployment of technologies, activities that often take considerable time to come to fruition.
- Underinvestment in long-horizon, technology-oriented projects by the private sector slows overall U.S. productivity growth rates, diminishing the relative standard of living of U.S. citizens.
- Underinvestment in long-horizon, technology-oriented projects also weakens specific U.S. companies in global competition with companies based in other nations, many of which appear to do a better job of investing for the long term.

- The loss of market share to foreign producers who are better at investing for long-horizon gains threatens U.S. economic security and further erodes the U.S. standard of living.

This argument is explicitly or implicitly offered as a partial explanation for U.S. economic problems in a host of studies of U.S. competitiveness performed over the past 15 years. The argument is widely accepted, and an obvious potential cause—a relatively high cost of capital in the United States—is often targeted in policy discussions. Despite its wide acceptance, the argument about the near-term orientation of U.S. companies has rarely been explored in any depth. Are short time horizons a ubiquitous phenomenon in U.S. industry, are they characteristic of only some industrial sectors, or of only small or large companies? What would explain a tendency toward short time horizons? Is there an identifiable link between lengthened planning and investment time horizons and improvement in corporate performance? Many very successful companies constantly seek to *shorten* their operating time horizons—by focusing research efforts more sharply on potentially profitable projects, by getting new products to market more quickly, by generating revenues from investments in plant and equipment as soon as possible, and by quickly and successfully instituting quality programs that increase profits. How can the assertion that U.S. companies are too short-sighted be reconciled with studies that show that the best-managed companies constantly strive to shorten the time frame of many activities? These questions, and similar related concerns, motivated the study.

THE APPROACH AND METHOD OF THE STUDY

In early 1990 the National Academy of Engineering formed the Committee on Time Horizons and Technology Investments (p. iii) to explore the determinants of investment time horizons, specifically with regard to the impact of time horizons on technology development and deployment, of both product and process technologies, by U.S. corporations. In addition to the committee's deliberations, the two-year exploration of time horizons and technology investments involved (1) two workshops at which members of the committee were joined by experts—from both academia and business—in finance, general management, employee and executive compensation, R&D and production management, and economics; and (2) a survey study of CEO's perceptions of the cost of capital (published as Appendix A to this report) commissioned by the committee and performed by Joseph Morone and Albert Paulson, members of the faculty of the business school of Rensselaer Polytechnic Institute.

Early in the process the committee discovered that there was no explicit, widely accepted definition of *time horizons*; nor was there much implicit agreement about the concept, its role in business activities, or its impact on U.S. economic performance. Chapter 2 makes its contribution simply by explict definition and discussion of "time horizons" and the relationship of time horizons to investments in technology development and deployment by companies.

Chapters 3, 4, and 5 explore and develop various influences on the time horizons and technology investment decisions of corporations. Chapter 3 addresses the roles of management and boards of directors in determining investment horizons and recommends strategies or approaches by which corporate executives can remove internal biases toward noncompetitive time horizons or mitigate the impact of external pressures to adopt noncompetitive time horizons.

Chapter 4 addresses the relationship of capital costs to investment time horizons. The chapter takes as its starting point the large and growing literature on national and corporate competitiveness, which often implies a direct, simple, and ironclad relationship between relatively high-average national capital costs, a perceived trend toward short-term trading in U.S. financial markets, and investment time horizons that are too short. The chapter reflects the committee's general finding that the relationship between capital costs and investment time horizons is complex, variable, and depends a great deal on the specific characteristics of the source of marginal capital, on the structure and practices of the company investing the marginal capital, and on the investment itself. The chapter makes some recommendations about private strategies and public policies to reduce harmful impacts that capital costs may have on time horizons for investments.

Chapter 5 examines the role of government investments and regulatory policies on corporate investment time horizons. The chapter is explicitly selective in dealing only with these two types of government influences on investment time horizons. Tax and fiscal policies (aspects of which are addressed in Chapter 4), and the specifics of regulation by the Securities and Exchange Commission, trade policy, antitrust policy, or intellectual property rights all are likely to affect the investment time horizons of U.S. companies. An in-depth treatment of such issues would stretch well beyond the expertise of the committee, which chose instead to focus its attention on two types of government influence on corporate time horizons that are not often explored or developed. Focusing on government investments and regulatory procedures, the chapter recommends actions and approaches through which government policymakers can productively lengthen the time horizons of private investment decisions in the United States.

The committee is grateful to the participants in the workshops for their numerous insights and contributions, and to Professors Morone and Paulson for their work, but the findings and recommendations in this report are based on the experience and consensus judgment of the committee. This statement, which is true of all Academy reports, is particularly important in this case because of the lack of empirical work or well-developed theory on investment time horizons; the study, by design, explored an undeveloped and poorly understood issue and the findings and recommendations, therefore, draw relatively more on the experience and insights of the committee members and relatively less on earlier empirical or theoretical work.

DO U.S. CORPORATE EXECUTIVES HAVE SHORT TIME HORIZONS?

Are some U.S. companies, or segments of industry systematically underinvesting in technology-related opportunities because of short-term decision making and investment bias? While there is considerable anecdotal evidence that U.S. firms behave in shortsighted ways, it is not an easy matter to generalize. In fact, there is remarkably consistent evidence that many U.S. firms have a rational spectrum of time horizons, from short to quite long. The most obvious is that some U.S. industrial sectors with high technical content and very long product cycles—for example, aerospace, pharmaceuticals, and chemicals—are highly effective international competitors. In addition, within almost every industry are examples of U.S. companies that perform exceedingly well and appear to have long planning and resource commitment time horizons. How can the performance of these industries and companies be reconciled with an assertion that U.S. companies are uniformly shortsighted?

The evidence that U.S. firms are shortsighted comes in primarily three forms. First, a large number of industry-specific competitiveness studies—cases of head-to-head competition between U.S. and foreign firms—have identified shortsighted behavior on the part of U.S. companies as one of the "fatal flaws" of the U.S. companies involved. The recent study of the MIT Commission on Industrial Productivity (Dertouzos et al., 1989) found that short time horizons contributed to the problems of the U.S. automobile, consumer electronics, machine tool, semiconductor, computer, copier, steel, and textile industries. The Commission was encouraged that the U.S. chemical and commercial aircraft industries were not apparently preoccupied with short-term goals. An earlier series of studies by committees of the National Academy of Engineering and the National Research Council found similar evidence of short-term behavior in several of the seven industries they studied (Steel and Hannay, 1985). In the summary of the findings of those studies, particular attention was drawn to the problems of steel and semiconductors because of the cyclic nature of their markets, abetted in the case of semiconductors by rapid obsolescence of product generations and of production equipment.

The failure of some U.S. firms—relative to their Japanese competitors—to invest adequately during downturns in demand is now part of a fairly standard story about competitive dynamics and the shortsightedness of U.S. executives. One well-known example is the semiconductor memories industry. Major U.S. companies chose to invest less in new products and new plant and equipment than foreign competitors during the time of slack demand. When demand increased, usually for the next generation of product, U.S. companies fell behind in their ability to respond while their competitors gained market share. As a result, the fate of some U.S. semiconductor firms was to lose market share coming out of every period of slack demand and eventually to leave the business. The short-term goals that executives choose to pursue—in particular the desire to earn "predicted" profits—seem to have hurt the companies in the long run. Use of a different goal—long-term profitable market share, for example—might have yielded different results.

MACROECONOMIC EVIDENCE: RELATIVE RATES OF INVESTMENT IN FIXED CAPITAL AND R&D

A second type of evidence that U.S. companies suffer from time horizons that are too short is the low relative levels of investment in long-lived assets by U.S. corporations. Between 1973 and 1985, manufacturing gross fixed capital formation as a share of manufacturing gross domestic product averaged 12.4 percent in the United States and 19.1 percent in Japan, a ratio of 1.5 in Japan's favor. From 1976 to 1988, investment in machinery and equipment in Japan varied from 14.9 percent to 20.6 percent of gross national product (GNP). In the United States it ranged from 7.5 to 9.0 percent. Rates of capital formation as a percentage of gross domestic product in other competitor nations—West Germany (before unification), France, the United Kingdom and Canada—were lower than in Japan but almost universally higher than in the United States. The last years of the 1980s were the most dramatic as Japanese investment in manufacturing increased by more than 25 percent between 1988 and 1989 while U.S. investment went up by only 9 percent.[1]

A third type of evidence that indicates short-term behavior on the part of U.S. companies is the low relative levels of investment in research and development—usually relatively risky investments not expected to pay off quickly. The United States actually leads the world's industrialized nations in terms of absolute expenditures on research and development, spending almost 2.5 times more than Japan. As a percentage of GNP, however, U.S.

[1]The most important international comparison would involve rates of net rather than gross capital formation, that is, the rate at which each nation is adding to its productive capital stock. There are, however, significant data problems even with measures of gross capital formation. The figures cited in this section probably do reflect significant differences in gross capital formation, but the exact amounts are subject to dispute because accounting practices in different countries define "investment" differently.

R&D expenditures are roughly equivalent to those of Japan, the United Kingdom, West Germany (before unification), and France. When defense-related R&D is removed, however, the results are quite different; the United States lags a full percentage point behind Japan and Germany in R&D spending as a percentage of GNP (in absolute terms, in 1987, the United States spent about $68 billion to Japan's $39 billion and Germany's $18 billion).

SUMMARY AND CONCLUSION

Although there is little agreement on the meaning of short time horizons, there is clear justification for concern over U.S. corporate investment time horizons. There are significant numbers of industries, or segments of industry, in which short-horizon behavior seems to be both the norm *and* a considerable source of competitive disadvantage, though a substantial number of U.S. companies and industries exhibit long-term behavior. In addition, there is macroeconomic evidence—low relative rates of investment in long-lived assets and in R&D—that appears to indicate a broad-based tendency toward short-term planning and performance criteria on the part of U.S. industry.

Evidence of short time horizons must, of course, be drawn from the recent U.S. economic history and from the assertion that U.S. companies have short time horizons is most closely related to concerns about U.S. industrial competitiveness in global markets. It is important to note, however, that the time horizons of U.S. private investment decisions will be important in a variety of contexts in the foreseeable future. In particular, U.S. businesses have seldom had to face such an uncertain and unstable future as they do today. Among the events defining the environment for business are the apparent end of the cold war and the impact of that change on U.S. policy and on defense industries; the demands of environmental protection and the requirements of capital for building and repairing infrastructures; and a relatively recently discovered pervasive weakness in the U.S. banking and insurance industries. The ability of U.S. companies to develop and maintain long-horizon investment plans—many of which must deal with the development or deployment of new technologies—through what is likely to be a period of substantial turmoil and restructuring will determine the economic prosperity and national security of the United States well into the twenty-first century.

2

Explaining Time Horizons and Technology Investments

The structure that a mature enterprise takes on at any point in time essentially represents the accumulation of a long series of prior resource allocation decisions. Opportunities to invest its limited resources arise continually, in a variety of ways, and must be acted upon by people throughout the organization. Their decisions regarding which opportunities to pursue and which to abandon, which aspects of the organization to strengthen and which to de-emphasize, and how much of their assets to devote to future rather than current needs, ultimately determine the firm's physical assets, human resources, technological capabilities, and overall competitiveness.

—*Robert H. Hayes, Steven C. Wheelwright, and Kim B. Clark,* Dynamic Manufacturing: Creating the Learning Organization *(New York, Free Press, 1988) p. 62.*

There is no single or standard definition of the term *time horizon* and no agreement on what business functions are affected by time horizons that are too short. What is clear is that time, as an element of planning, decision making, and execution, is a crucial aspect of competitive performance in a number of industry sectors. Examples of the role of time in company activities include

- The time required to commercialize a new product or service that depends on the development and deployment of new technology
- The planning time frames (operating, business, and strategic) for which a company develops actions it chooses to pursue

- The time needed to build critical skill bases and teams, or to develop or deploy long-lived assets needed to improve company productivity
- The expected time between investment in development of a new technology and payoff
- The time it takes for a new market to develop and become saturated
- The length of time ahead that an organization can plan because of uncertainties affecting forecasts (procurement cycles, legal changes, or regulatory practices) for the industry
- The time it takes for a competitor to copy a product and get that product to the market
- The time scale embedded in the employee incentive and reward system

This list makes clear that every corporation operates with a host of different time horizons for its activities; companies must balance a range of different time-dependent business activities. In addition, companies in different industries obviously face different time horizons as a function of different economic, technological, market characteristics, and competitive conditions. Figure 1 shows the variation in company options through the

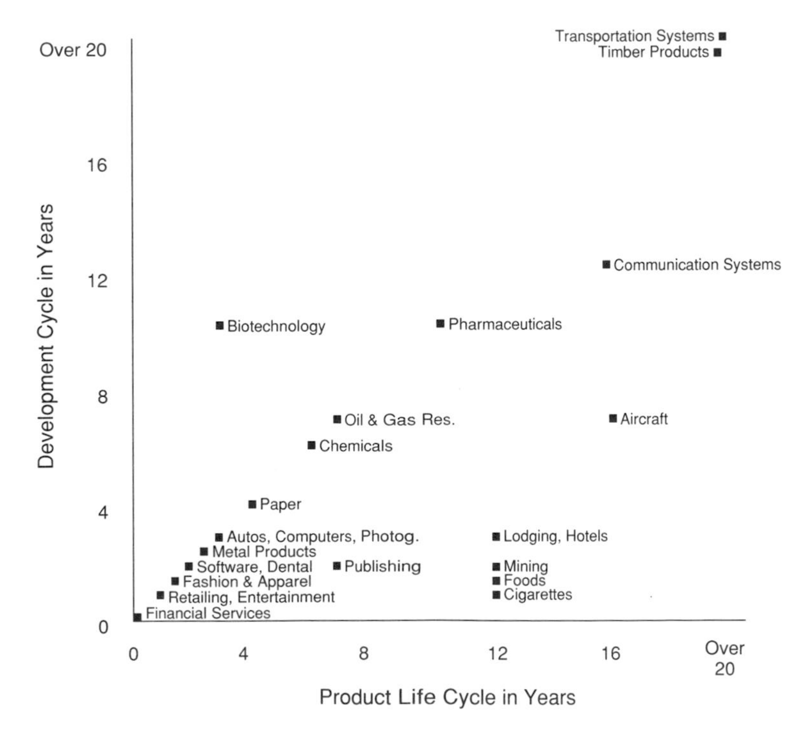

FIGURE 1 Typical time horizons by industry. Graph adapted by J. B. Quinn, from a concept introduced in a seminar by W. H. Davidson, University of Southern California, Spring 1986.

wide dispersion, by industry, of both the development times of new products and the market life of products.

The position of different products in Figure 1 shows how two important operational time constants—time to develop and market new products and market life of products—vary by industry. The implication of these variations in industry-specific time cycles is that there will be substantial "natural" variation among industries in many time-dependent business matters. Industry norms for research and development funding levels, development investment per product cycle, plant and equipment investment life, new product pricing strategies, employee reward systems, and competitive strategies are all affected by industry-specific timing factors.

Industry-specific variation in time-dependent business matters illustrates an important point about time horizons: that individual company management and governance practices play a fundamental role in determining time horizons. Companies in industries with long product or market development cycle times—pharmaceuticals or airframes, for example—must have relatively long investment horizons. Stable, successful companies in longer product cycle businesses—and there are many—are proof that effective management can collect and organize financial, human, and technological resources for competitive commercial activities with payback far in the future. This conclusion is buttressed by the fact that, within a given industry, it is possible to find companies with different time horizons and different levels of success. Companies in a single industry face a similar competitive environment, yet some are able to compete much more effectively than others. Such companies have different methods of managing, different time horizons and, consequently it seems, different levels of performance.

WHAT IS "NEAR-TERM ORIENTED" MANAGEMENT AND GOVERNANCE?

Management practices and decisions, in concert with governance structures and practices, play a large role in determining the time horizons that a company exhibits. The willingness and the ability of managers to address the longer-term future of a company are especially critical to a company operating in a technologically dynamic business. The same is true of the demands on a corporate board or on active venture investors; if the governance structure of a company is biased toward short-term return, it will be almost impossible, no matter what the external influences, for the company to develop and deploy new commercial technologies.

One effective way to characterize management's time horizons relies on time-to-break-even charts (also called return maps—see House and Price, 1991). Time-to-break-even charts show cumulative cash flow plotted over the life of a project. Figure 2 shows hypothetical time plots of the sum of

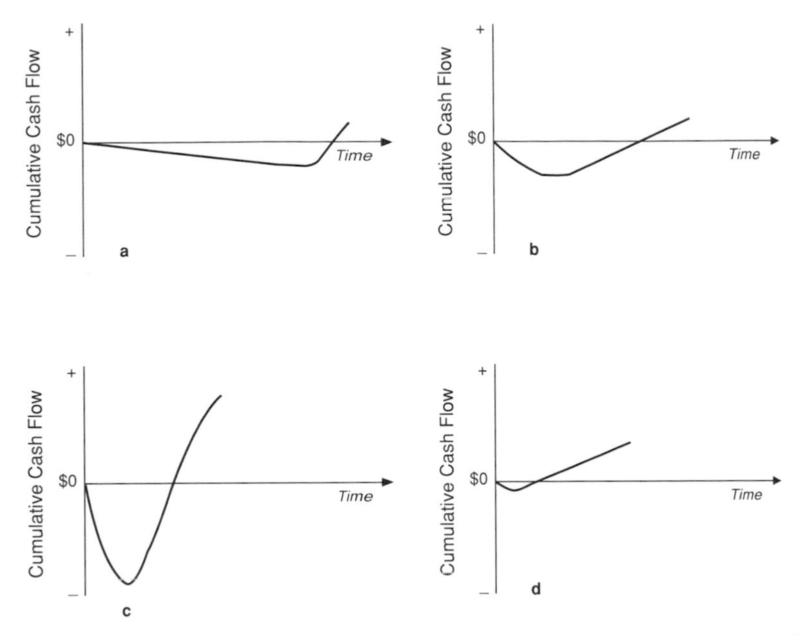

FIGURE 2 Graphs illustrate hypothetical cash flows for (a) a project with a long research phase, (b) a typical development project, (c) a major investment in new plant and equipment, and (d) instituting statistical process control in a factory.

forecasted (or actual) expenditures and revenues in carrying out different types of projects:[2]

- Figure 2a shows the cash flow of a project involving a protracted period of research—slowly accumulating low-level negative cash flow with a substantial payoff occurring far in the future.
- Figure 2b shows a prototyping, development, and marketing project with rapidly accumulating initial investment expense and a relatively quickly occurring and high-volume revenue stream.
- Figure 2c shows the cash flow of an investment in new production equipment—a very expensive investment with rapidly accruing and large cost savings.
- Figure 2d shows the costs, timing, and returns of instituting statistical process control in a factory—low initial expense bringing on a rapid but small incremental revenue (cost saving).

[2]The proper formulation of these curves should be sensitive to methods of cost allocation for the whole profit center. If the accounting system allocates a portion of the profit center's fixed costs to the project's negative cash flow (thereby reducing costs and increasing profits on other profit center activities), then the curve should describe the net effect of the project on the whole profit center cash flow.

These figures illustrate a portion of the portfolio of investment options that virtually all companies face; with limited resources for investment in any period of time, a company must make choices among the various options, thereby choosing a mix of expected expenditures and returns. In most cases it is good business practice for managers to establish a balanced portfolio of investments, one that includes cash users and cash generators and both high- and low-risk investments for both the short and the long term.

The range of investment options characterized in Figure 2 can be used to illustrate the arguments that fault U.S. industrial management for being shortsighted. In general, the concept of shortsighted management revolves around a perceived preference (for whatever reason) on the part of corporate managements to invest in activities in which the break-even point occurs relatively quickly after an investment is made. The assertion is that the portfolio of investments that U.S. companies choose is too heavy with those that are likely to yield returns in the near future. The assumption is that U.S. managements are passing up options that are—in some long-term sense—more valuable to the company (and, by implication, to the nation) than the short-horizon projects that companies do pursue.

While we have chosen to illustrate the argument about short-sighted U.S. corporate management using a project analysis and management tool (time-to-break-even graphs), it should be clear that the argument applies broadly to management decisions. For example, a company's management may pay particular attention to operating measures, such as profitability ratios (e.g., return on total assets employed) or activity ratios (e.g., inventory turnover). These tools are primarily useful for comparing an operation's performance relative to others in the same industry or to itself at another time. From the perspective provided by these tools, the short-time horizon argument revolves around whether or not U.S. executives have adequate patience to be competitive. Managers with short time horizons will favor investments in already performing assets (a business line or factory) over assets that may have greater long-term potential but reduce near-term earnings.

In this context, it is easy to see how uncertainty and the related investment risk contribute to shortening time horizons. A project that takes longer to come to fruition is exposed to competitors, faulty cost and schedule estimates, changes in the economic or regulatory environment, or failures in company performance for a longer period of time. The longer an investment takes to develop, the longer it is exposed to the possibility that key personnel, including corporate planners and decision makers, will lose interest, changes jobs, or retire. Referring to a time-to-break-even chart, the longer the expected project profile, the more uncertainty there is about the validity of the forecast of when the curve will turn up, if it will turn up at all, and how fast it will rise. The same logic applies in the case of operating ratios—the longer the expected time for improvement, the less certain a manager is that the ratios can be made to improve adequately.

Much marketplace uncertainty (and the risk it creates) is the natural countervailing force to long-term planning and investing; decisions that generate a quick, more certain, payoff enjoy a genuine advantage over projects with higher long-term potential but higher risk. A bias in favor of activities with more certain return (often shorter-term investments) is a desirable trait for many managers. Most important, it is a trait that becomes more desirable with increasing uncertainty in the economic and competitive environment. Constantly fluctuating tax or regulatory policy, rapidly changing currency exchange rates, significant uncertainty about market acceptance of a new technology, or a cadre of well-funded, aggressive competitors can all increase the apparent value of a manager who focuses on the immediate future.

THE MECHANICS OF CAPITAL COSTS, RISK, AND THE SPECIAL CHARACTERISTICS OF INVESTMENTS IN TECHNOLOGY DEVELOPMENTS

The cost of capital—the required expected return derived from uncertain future cash flows—can dramatically affect the time profile of investment decisions. Although determining the real cost of capital for a company requires a complex estimation based on amounts and handling of debt, equity, and retained earnings (and, depending on the measure, tax and depreciation effects), it is easy to illustrate the impact of more expensive capital on the portfolio of investment options a company faces. Figure 3 shows a time-to-

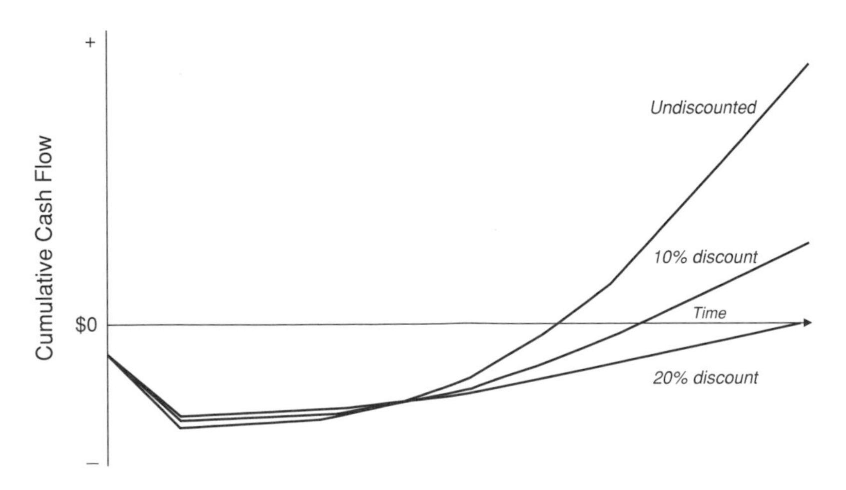

FIGURE 3 An estimated time-to-break-even cash flow curve calculated using three different expected rates of return.

break-even cash flow curve calculated using three different expected rates of return on capital: (1) undiscounted cash flow; (2) a 10 percent discount rate; and (3) a 20 percent discount rate. The break-even point shifts further and further into the future with every increase in the cost of capital. This effect applies across the board to all project options and to return calculations on all assets. Therefore, a manager who pays more for funds for investment faces a set of investment options that take more time and risk to produce an adequate return. The higher the cost of capital the more tightly constrained a manager is to select those options with rapid payback.

Perspectives on the Cost of Capital

From the perspective of financial investors, the required expected return from an investment is very sensitive to risk. Figure 4 shows a standard model of the increase in return demanded for increasing risk by U.S. financial markets—the capital market line. Any individual financial investment, or

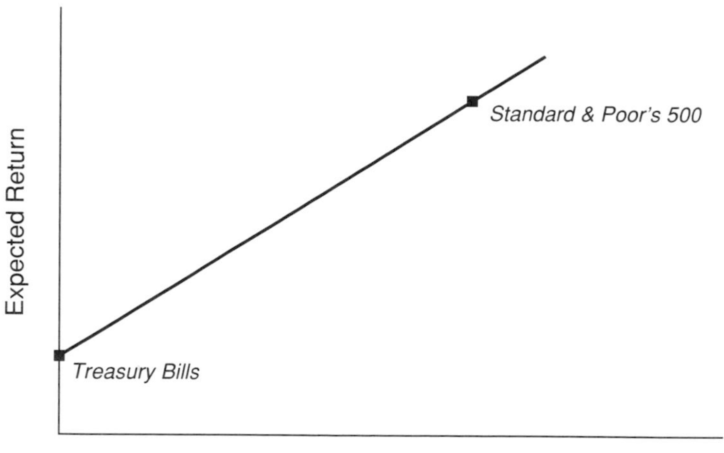

FIGURE 4 The capital market line. Over the past 65 years, the real return on Treasury bills (generally considered a safe asset) has been 0.5 percent, whereas the arithmetic average of the real rate of return on the Standard and Poors 500 has been 8.8 percent per year. A physical investment (or a human capital investment) whose riskiness is comparable to the Standard and Poor's 500 would have to be competitive on an after-tax basis with the 8.8 percent return on the stock market, not the 0.5 percent return on Treasury bills.
SOURCE: Shoven (1990, p. 5). Reprinted with permission.

portfolio of investments, can be characterized by a point on a plot of return against risk, not necessarily on the capital market line. The mean and standard deviation combinations on this capital market line characterize the demands of the financial market for returns at different levels of risk (Shoven, 1990). The capital market line characterizes and explains the standard definition of the cost of capital as the demands of investors: the required expected return derived from uncertain future cash flows.

An alternative use of the concept of the cost of capital is best described from the perspective of owners and managers of nonfinancial companies who deploy capital. From their perspective the required expected return by investors is the cost of funds. The cost of capital is the pretax rate of return necessary to pay any taxes *and* the cost of funds. This definition allows decision making on the basis of discounted cash flows or net present value estimates, calculations that discount cash flows by a firm's cost of capital. A related concept is that of an "investment hurdle rate"—the discount rate that is part of decision-making rules or procedures within a company. The creation of investment hurdle rates often starts with the firm's pretax cost of capital and adds premiums for risk to establish the required expected return on a specific investment or type of investment.

Both the market rate cost of capital and internal investment hurdle rates are important, and they are intimately related. The first definition, which relates to the cost of funds, reflects the financial market's perception of the risk of investing in a company. In large and diversified companies an investor buys a risk/return package that reflects a bundle of company activities, some risky (R&D, for example) and some less risky. The second, as an internal decision-making tool for resource allocation, affects marginal decisions by managers in selection among investment opportunities. For example, should a company spend more on R&D at the expense of upgrading existing plant and equipment? Management decisions—made on the basis of the discounted cash flow projections—will affect the bundle of activities in a company and thereby ultimately affect the company's market cost of capital. Sources of uncertainty (and therefore risk) also affect both types of capital costs.

Although it is widely recognized that risk affects required expected return both within companies and in financial markets, it is less well understood that the character of commercial technological advance poses special problems both for investors (financial markets) and for managers working with investment hurdle rates as a guide in making difficult investment allocation decisions. The demands of developing, deploying, and managing complex product and process technologies create a type of risk that is not necessarily well handled either by financial markets or by managers making investment decisions.

Financial Markets and Technology Investments

With regard to financial markets, the relatively long time periods necessary for technology investments to come to fruition raises the issue of "patient" capital. The point is clearly illustrated by an example of a new company developing and marketing a single product. It is not unusual for product development to take three years, production design two years, and initial market development, concurrent with production tooling and buildup, another two years. It may take an additional five years for the company to show its full potential. In addition, for companies working with newer technology, the ability to react quickly to changes in market demand, consumer preference, and competitor capabilities depends on technical capabilities that usually must be developed and nurtured over a period of several years.

During the substantial time between start-up and a significant revenue stream (the time it takes to prove a product or service in the market) an investment in a technology-based company can be both intangible and highly illiquid. Many investments in technology are intangible in the sense that they are expenditures of funds on learning how a product should be designed or produced, or how a particular market needs to be developed. Such learning investments—in contrast to real estate, capital equipment, or a license to manufacture—are an intangible asset not easily sold or used as collateral.[3] As a result an individual investor's exit from a technological venture can be hampered by a very thin market for an interest in an unproven product, and failure (often in the form of bankruptcy) may leave little residual value for any equity investor.

During later stages of a successful new company's growth, an investment becomes more liquid, but an investor who wants to exit may pay a substantial penalty for getting out before the investment is mature. The intangibility and illiquidity of the investment apply to even the most successful technology investments (high annual rates of appreciation if calculated over a long period). In other words, even in technological ventures with good long-term prospects, the characteristics of technological development demand *patient* capital—investors willing to take their return mostly through long-term appreciation.

The economy has developed a variety of mechanisms to provide patient capital to support new commercial technologies. Financial markets do allow investors with a preference for high-risk, high-potential, long-term payout investments to get access to new, potentially successful technology-based

[3]Tangible and intangible investments are also treated very differently for accounting purposes. Tangible investments are capitalized, whereas intangible investments, such as training or R&D, are usually expensed in the current year. For a discussion of tangible and intangible investments, see Hatsopoulos and Brooks, 1986.

companies; the technology-oriented segment of the U.S. venture capital industry specifically matches that set of investor preferences to opportunities, and the United States has a highly developed and smoothly functioning high-technology venture capital market. More important, larger, multiproduct corporations allow equity investors to buy a bundle of corporate activities, some of which may be risky new product and market development activities. The fact that technology developments are bundled with less uncertain activities (i.e., existing, successful product lines) provides investors with liquidity and can create a company-specific pool of patient capital for technological risk taking.

Although these mechanisms for providing patient capital do exist, they are not necessarily optimal or even adequate. It is often argued that investor expectations for risk, liquidity, and short-term return from equity holdings in public companies—the cost of publicly traded equity capital—inhibit risk taking, such as technology-oriented, long-horizon investing. It is also true that formal venture capital markets serve only very specialized high-growth-rate opportunities in selected industries; technology-based start-ups can face feast or famine in trying to find venture capital because of relatively thin and uneven investor experience and interest. Finally, some technologically dynamic companies face considerable risk-related problems in simply obtaining loans for growth or modernization. Small companies (under $20 million a year in sales) and technological risk-takers—depending on how their industry is perceived and the state of the economy—may have no effective access to capital despite reasonably good company prospects.

Investment Hurdle Rates and Technology Investments

With regard to investment hurdle rates in management decision making, the primary issues related to technology investments revolve around ways in which management assesses and handles technological or market uncertainty in investment decision making. In the same way that managers and owners make investment decisions with different projected time-to-break-even profiles (see Figure 2), it is also true that managers and owners make simultaneous investments of varying riskiness. As discussed above, these two characteristics of investments—time profile and riskiness—are often correlated, though not perfectly. An investment in developing a new product for a highly competitive market is probably more risky than investing in an upgrade of an existing successful product. The two investments will have different time profiles, different degrees of risk, and can be directly compared, formally or casually, by using a higher investment hurdle rate for the riskier investment.

Table 1 lists factors that will, from the perspective of a manager or owner, increase or decrease the risk of an investment in developing or deploying a new product or process and, as such, affect the appropriate

TABLE 1 Factors that Increase and Decrease Risk Associated with Technology Investments

Factors Increasing Risk	Factors Decreasing Risk
Low experience in the market with the product or service	Expansion from existing strengths
Strong competitors	No dominant competitors
Technological uncertainty	Government technology support or steps to create the market
Environmental uncertainty	Stable standards—environmental, technological, social, etc.
High potential product liability (medical products, nuclear, toxics, etc.)	Relevant government infrastructure
Changing standards	External investment partners (cooperative venture)
Less than strong confidence in internal capabilities	"Safe harbors" from product liability for certain products (vaccines, defense products, etc.)
Restricted market access, especially in worldwide markets	Access to foreign or government technology or other external sources
Little protection for intellectual property	Protection for intellectual property

investment hurdle rate for the investment. Managers or owners who make resource allocation decisions in technologically dynamic companies are constantly challenged to weigh such uncertainties in investment decision making. Whether or not decision makers formally reduce risk factors to a premium added to an investment hurdle rate, it is obvious that the relevant *internal* cost of capital (investment hurdle rate) for technology investments is specific to the investment. In this context, it becomes clear that a company's ability to manage and commercialize technology effectively will determine its time horizon for technology investments.

A company that has a deep, reliable competence in commercial development and application of a new technology will take less risk in any particular technological investment than a company that is technologically weak. As a result, a company that is effective at technology management and application will attach a lower-risk premium to a technology investment, allowing it a longer time horizon. Such impacts may be more important to time horizons than economywide conditions such as the average cost of funds or even the marginal cost of funds for a particular firm.

In closing out the discussion of technology investments, it is important to note that the arguments relating technology, management, and investment do not apply exclusively or even predominantly to new end-user products based on new technologies. The same concerns about investment and time horizons apply to new production technologies and technology-based services, in some cases even more strongly than in new product development cases. For example, some production technologies are "leverage" technologies—crucial sources of production efficiency and competitive advantage but often not a large portion of the cost of end products. Such technologies can be expensive to develop and benefit mostly those companies with substantial long-term presence in the end-product market.

DETERMINANTS OF COMPANY INVESTMENT TIME HORIZONS

As discussed earlier, there is considerable variation in industry-specific time constraints for such things as market life and product development cycle. Some important determinants of time horizons obviously operate for entire industries or industry segments. Prevailing economic conditions—such as the cost of available funds or the rate of growth of consumer purchasing power—or the competitive or product cycle status of the industry can set tight bounds on the options companies can pursue. The legal system can also have a great deal to do with time horizons in an industry. Such is the case in the pharmaceutical industry, where the length of time for drug clearance and patent protection is directly related to return on R&D investment. In yet another set of circumstances, for example when a new market is growing rapidly (where there is significant pent-up demand), the time horizons of companies will be determined less by prevailing economic conditions or legal concerns and more by the speed of competitors.

Another set of important influences on time horizons is highly company-specific and dependent on management and governance decisions. Particularly important is the company's competitive position and trends in the company's market share and profitability; a company that is widely perceived to be losing market share or is less profitable than competitors is likely to focus sharply on the immediate future—the time frame in which its survival will be determined. Additionally, the time horizons of companies are affected by operating routines and practices such as methods of selecting projects, production capabilities, marketing abilities, incentive systems, methods of strategy development, career systems, and methods of raising capital.

Company size and growth rates are also important. For example, small companies and large companies have different needs and limitations with respect to capital availability, capital costs, and sources of new capital. Capital that appears expensive to large companies may simply be unavailable at any price to small companies. The challenges growing companies face also differ from those of mature organizations. Smaller, rapidly growing companies may have greater flexibility in their planning because they lack the organizational inertia of bigger, more mature organizations. Such companies, however, may also be less stable and therefore relegated to short-term planning aimed at keeping the company in business.

In other words, all companies operate in a complex financial, legal, and competitive environment, and there is clearly no single determinant of corporate investment decision making and planning time horizons.

SUMMARY AND CONCLUSION

The common presentation of time horizons in U.S. companies—that U.S. executives are generally near-term oriented and pursue only short-term goals—is too simplistic. The situation with regard to time horizons and technology investments is complex, as time horizons for technology investments should, and do, vary widely by industry, product, and business activity. It is clear that short time horizons are not a universal problem for U.S. companies; there are a number of successful U.S. companies operating in industries that require relatively long time horizons for investments.

Near-term orientation in company behavior can be understood as a preference, subtle or explicit, for a portfolio of investments that are likely to yield returns in the near future. Such preferences are in many cases rational, as technological and marketplace uncertainty, and the investment risk they create, are natural countervailing forces to long-term planning and investing. This link between risk and short time horizons is quite explicit in the role that capital costs and investment hurdle rates play in investment decision making in companies; the more risky the project or venture the more likely it is that both financial markets and internal management decision making processes will require a higher expected return.

The relationship between risk and investment time horizons is particularly important with regard to investments in the development and deployment of new product or process technologies. Investments in technology-dependent ventures may create largely intangible assets, they may be illiquid for long periods of time as projects can be slow to mature, and they are exposed to both normal business risk and technology-related uncertainty. As a result, technology investments often carry a substantial (formal or informal) risk premium. Although some of the risk is irreducible a substantial portion reflects questions about the ability of a company to bring a competitive new product to market or to introduce a substantial process innovation. This implies that short time horizons in technology-dependent investments can be caused by the inability of companies to manage technology effectively; companies with deep and genuine competence in commercial application of technology will have a distinct advantage in adopting longer time horizons for technology investments because they can reduce the risk of their investments.

Important aspects of any company's options, practices, and time horizons are created by a diverse and interactive set of factors, some of which are clearly the prerogative of management and some of which are set external to the company. The specific competitive status of a company, uncertainty, and the abilities of a company's board and executive managers to deal with uncertainty, will affect the time horizon of specific decisions. Also, the expectations of investors (the cost and patience of capital) will interact with the financial structure and investment practices of a company to affect the time horizons of a company's decisions. Finally, the design and implementation of government policy can affect the time horizons of companies. This diversity of influences on corporate time horizons implies that no single actor can unilaterally lengthen investment time horizons. The federal government, boards of directors, and company management will need to act, both separately and together, if U.S. technology investment time horizons are to be lengthened.

3

Company Time Horizons and Technology Investments: The Roles of Corporate Governance and Management

In cases where the unprecedented developments of recent years have contributed to . . . long-term perspectives by motivating managers and financiers to define and implement long-term plans for restoring, maintaining, and improving organizational capabilities, they have helped to make enterprises, industries, and nations more competitive and profitable. But where these developments have encouraged short-term gains—where decisions and actions have been motivated by the desire to obtain high current dividends or profits based solely on the transactions involved in the buying and selling of companies—at the expense of maintaining long-term capabilities and profits, they appear to have reduced and even destroyed capabilities essential to complete profitably in national and international markets.

—*Alfred D. Chandler, Jr.,* Scale and Scope: The Dynamics of Industrial Capitalism *(Cambridge, Mass., The Belknap Press of Harvard University Press, 1990), p. 627.*

The volume and pace of commercial technological development are increasing in many industries. New products and processes are being introduced more frequently and, from the perspective of an individual company, there are narrower and narrower windows for product introductions. Corporate product development efforts are getting increasingly more sophisticated at surveying and incorporating customer preferences. This pursuit of niche markets demands faster product changes and more flexible marketing departments, design departments, and manufacturing processes. The implication is that organizations must develop and improve their products and processes with increasing rapidity. Coupled with the immediate need to fix current operational and competitive problems, and the need to meet aggressive financial targets, this can drive managers to focus on the short term. A critical concern for both small and large companies in this situation is the need to excel at product and process renewal, and the development of revolutionary products and processes, without sacrificing the development of long-term organizational competencies.

The role of management and corporate governance in effecting successful investment in technology is evident in any examination of typical company operations and can be illustrated with another hypothetical time-to-break-even graph—Figure 5. The vertical arrows indicate common events in the life of a corporation, many of which will directly affect the viability of a technology-intensive project. Most of these events are at least partially under the control of corporate management or a function of governance decisions. A corporation that avoids or manages such events well will exhibit noticeably more constancy of purpose and, probably, longer time horizons than one that does not. The purpose here is not to argue that external influences do not affect corporate time horizons but rather to recognize that management practices can either reinforce or run counter to external influences on time horizons. If the external environment pressures

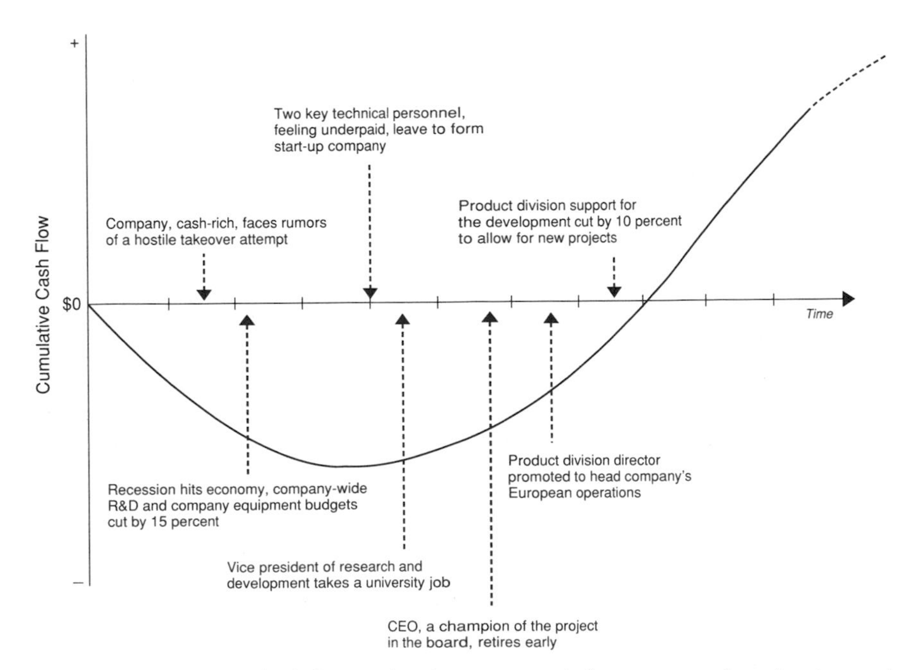

FIGURE 5 Hypothetical time-to-break-even graph for a new product development and common events in the life of a corporation.

a company toward inappropriately short-term investment decisions, it is management's responsibility to work against such pressures, to mitigate the impulse to focus too narrowly on short-term accomplishment.

Of course, even the most effective management will not always be successful in overcoming external influences. A company can develop a deeply rooted, short-horizon culture, a culture in which both explicit and implicit criteria for performance evolve to reward and thereby reinforce short-term behavior. For example, a corporation that faces a cost of capital that is high relative to competitors may reward managers who minimize capital equipment, training, and educational investments (longer horizon payback) and who select low-risk, short-term projects. Once this culture is embedded in an organization, even if the real relative cost of capital is significantly reduced, the behavior of personnel in the organization will not change perceptibly in the short term. Employees who have risen in the organization will have done so because it is natural for them to be risk averse and short-term oriented. Even if there are striking changes in the cost of capital, there will not be rapid changes in corporate behavior. This persistence of company behavior in spite of changing external conditions is a familiar phenomenon—one encountered by any manager who has tried to redirect an organization.

Among the factors important to the time horizons of a company's investments and largely under the control of company governance or management are the following:

- Investment decision-making and planning methods
- Incentives and reward plans for management
- Operational and project management techniques
- Career paths and patterns for employees
- Corporate financial structure and practices

This section describes major management and governance practices that influence the time horizon a company exhibits. The role of management and corporate governance in capitalization (debt/equity structure, decisions about how and when to raise capital, methods of obtaining investment capital without going to the financial markets) is dealt with only briefly in this section. They are discussed more in Chapter 4, in the context of capital costs.

THE IMPORTANCE OF GOVERNANCE: THE ROLES OF BOARDS OF DIRECTORS

Although the forms of corporate governance structures are circumscribed by law, the variation in practice is substantial. At one end of the spectrum is a small, venture-capital-backed company, the board of directors of which may consist of one or two venture capitalists (representing a pool of investors), the founder, and a senior manager of the company. In such situations, most investors are intimately involved in, and knowledgeable about, the operation of the company and are either personally interested in its success or directly and immediately responsible to a group of investors. Insiders and outsiders (the venture capitalists) are likely to share a common set of interests as owners; if everyone attends the board meeting, 100 percent of the outstanding equity is well represented at the table.

At the other end of the spectrum are large, public companies with boards of 15 to 20 members or more. With rare exception, no individual director will personally hold more than a few percent of the company's outstanding equity, and most hold much less than 1 percent. The degree to which directors effectively and actively represent shareholder interests is often low. The sheer size and complexity of large public companies mean that, in many cases, inside directors are the only individuals deeply knowledgeable about the company operations. Additionally, financial market structures and practices can minimize meaningful communication between the shareholders and directors. An increasingly important segment of shareholders for large public companies includes institutions (such as pension funds, mutual funds, insurance companies, and trusts) that have professional investment managers who themselves represent others, who often know little about the business the company is in, and who are far more likely to trade out of a company that is performing poorly than to try to improve performance through corporate governance mechanisms. Large-scale, pooled indexed funds are the extreme example as they trade bundles of stocks—each one of which is effectively ignored in buying or selling.

The time horizons a company exhibits are a reflection of corporate goals, directions, and strategies—issues that are the responsibility of corporate governance structures. As such, boards of directors are intimately involved with, and bear substantial responsibility for, directing corporate time horizons. Boards and the top management they select are the people who consistently have access to the type of information that facilitates a high-level, long-term view; they are uniquely responsible for the company's future. Thus, if a public company's performance is weak because of shortsighted investment behavior, it is a failure of its board of directors.

Boards of Directors' Choices and Time Horizons

Ultimately the board of directors is responsible for the time horizons and performance of a company. In many situations this responsibility manifests itself in the board's role in balancing two countervailing financial pressures: the pressure to invest in the activities that secure the survival and future prosperity of the company and the pressure to pay a return to current stockholders. In a technologically dynamic business, where survival and future prosperity may depend on substantial distant-horizon investment in R&D or new plant and equipment, the balance can be especially difficult to maintain. The same is true in high-growth (often newer technology) sectors where demands for new investment are high relative to industry sales. The ability of a board of directors to work with senior management to negotiate such tensions will play a fundamental role in determining the time horizons and performance of the company.

Unfortunately, few boards of directors of large public companies seem to have the inclination, power, time, or competence to do more than review and approve the proposals of company executive management and the CEO in particular. As long as the proposals are well conceived and in the shareholders' interests, boards of directors appear to perform their functions satisfactorily. However, although they nominally have the authority to exert far more influence than they normally do over company policies, plans, and operations, most boards of directors seem content to advise and consent, reviewing and discussing and, ultimately, approving the propositions that are presented to them by the CEO.

Specifically with regard to time horizons, the board of directors (with the help of senior management) importantly influences the time horizon of a company through the selection and development of senior management. Boards of directors have as a principal responsibility the choice and continued appraisal of the chief executive officers and the development and balancing of senior company management. To be effective at governing a corporation, a board of directors should be attentive to the importance of balance within the senior management team. The following items constitute a short list of concerns about balance in senior management in relation to time horizons: (1) the age distribution of senior management; (2) the degree to which there are high-quality, identifiable champions for the initiatives that should mature into a company's core businesses in the next decade; and (3) the balance between members of the senior management team who are focusing on near-term problems and those who are focusing on the long-term future of the company.

A board of directors is responsible also for designing compensation packages for senior management in a way that is in the best interest of the company. As straightforward as this responsibility appears, it is not uncommon for a board of directors to put together or accept a compensation package that is (1) largely insensitive to corporate profitability or (2) very sensitive to the value of stock prices at the time of retirement or departure (Jensen and Murphy, 1990). Boards of directors should link compensation packages for senior executives to their performance in developing and implementing plans for the long-term prosperity of the company. A compensation package linked to stock price at the time of retirement can provide a strong incentive, depending on the age of the individual and the vesting period, but can also work against longer time horizons if it provides a senior manager with an incentive to "cash out" at the highest short-term stock price without regard to the future of the company.

The actions of boards of directors are crucial to the time horizons of a company and, therefore, so are the methods by which directors are selected, compensated, and removed.

Compensation and Selection of Board Members

The weak performance of some boards of directors in serving the interests of shareholders seems to arise from (1) a lack of information that has not been filtered through the corporate executive office and presented to the board; (2) insufficient time dedicated to reviewing, evaluating, and understanding the information that is available to the board; and (3) insufficient expertise to understand and formulate well-considered opinions based on the information. These common problems of members of boards of directors arise, in part, from the common origins of many directors as sitting or recently retired CEOs, individuals active in public life, or prominent academics. These individuals clearly have deep and useful expertise in many areas, but they may have too little time, incentive, or experience in the relevant industries to be effective board members.

Among the most common suggestions for improving the performance of the boards of large public companies is to provide structures and incentives to drive them to behave more like the boards of well-governed, small, venture-stage companies. In general, this implies that directors need to be more responsive to shareholder interests, less captured by senior company management, and generally more involved in review and resource allocation (Jacobs, 1991; Johnson, 1990; Patton and Baker, 1987). One tool for moving in this direction is the method of compensation for individuals who serve as directors.

Stock ownership valued at several times annual board compensation might better align director's interests with the interests of shareholders; at present, there is no legal requirement for directors to own stock in the corporation on whose boards they sit. Corporations could move to increase the financial stake that their outside directors have in the corporation by requiring that they own shares at least equal in value to a specified multiple of their annual fees as directors. The intent is to link board member compensation as directly as possible to long-term performance in stock price. For new directors, this could be accomplished by paying them only in stock for the first years until the requirement is met. An additional incentive for long-term thinking on the part of directors would be to offer some significant portion of directors' compensation in stock options that can be exercised only after several years.

It is important to note that directors' compensation schemes are not a cure-all for the many perceived ills of boards of directors of public companies. It is easy to argue, given the time commitment and legal exposure of directors, that the vast majority of outside directors of large public companies do not sit on boards because of the compensation. Directors sit on boards for a variety of reasons: as a favor to the chief executive officer, for prestige, to gain experience in another industry, or to gain correlated experience in the same industry. This raises a second issue regarding ways in which directors are selected and removed. Directors and executives in larger U.S. companies tend to come from a common pool of business leaders and as such, it can be argued, best represent the concerns of professional managers rather than investors. One approach for avoiding this situation is for companies to establish a nominating committee of the board consisting entirely of outside directors with sufficient staff support to permit a thorough and competent job to be done.

THE PREROGATIVES OF MANAGEMENT AND CORPORATE TIME HORIZONS

The direction and operation of a corporation is a complex and diverse task often requiring a wide array of managers; companies have top management, sector management, group management, division management, management staffs, functional management, and program management. Some of these managers have immediate-focus, operational positions. Others—usually the more senior management (by whatever title)—have responsibility for strategic direction and significant resource allocation decisions. This section discusses the ways in which senior management affects a company's time horizons. Most of the important roles of senior management with regard to time horizons can be grouped into three categories: (1) constancy of purpose coupled with flexibility in the development and execution of corporate strategy; (2) design and implementation of career development systems and compensation schemes; and (3) design and choice of decision-making methods and measurement tools. Each of these categories is discussed below.

The Development and Execution of Corporate Strategy

Investment, planning, and operational time horizons are an integral part of strategic planning and execution. An indicator of the degree to which time horizons are adequately treated is the balance of time horizons—from immediate necessary actions, through investments expected to bear fruit in two to five years, to actions expected to ensure the performance of the company five to ten years out. Many companies use a portfolio approach to ensure that there are both long- and short-term projects as well as high- and low-risk projects. The length of the appropriate set of time horizons is, as noted earlier in this report, dependent on the business sector.

Within this framework, however, managements can vary their mix of long- and short-term projects significantly. Within the same high-technology industry, for example, companies with a basic research strategy will tend to operate with time horizons seven to nine years ahead, surround their discoveries with patents, and use their exclusive positions to increase profits. Others may wait until initial market risks have been taken by others, then move rapidly into the fray. Still others may use a mixed strategy of protecting certain "core lines" of the business with long-term depth and technology, while moving quickly and interactively in other marketplaces using their established distribution capabilities.

Many companies follow such mixed strategies. When they do so explicitly, they may consciously override direct project-by-project present-value comparisons between the different segments of their business and invest in the longer-term segments based on a judgment that these projects will contribute higher payoffs in the long run. The nearer long-term projects move toward the basic research spectrum, the greater this element of subjective judgment must be. A conscious portfolio strategy to defend long-term, uncertain investments in support of specific strategic thrusts can be a rational part of maximizing long-term corporate returns.

Managements are virtually always forced to balance two competing pressures in the execution of a strategy. On the one hand, constancy of purpose in management is important for long-term projects; clearly if top management chooses a strategy for the organization but does not remain firm in its resolve to pursue the goal, managers in the organization will be unable to make appropriate decisions and both the long-term and the short-term health of the company may be jeopardized. On the other hand, management needs constantly to weigh the potentials for a successful outcome from following a given strategy against the potential for losses due to blind pursuit of an elusive goal. The proper use of decision gates—explicitly planned "what if" points at which investment plans are reevaluated—can help management avoid vacillation without incurring unnecessary losses in the execution of strategy for the company.

This tension between constancy and flexibility is primary, but only one of several tensions in which management judgment and execution can make the difference between good performance and failure in management of a technology-intensive business. The following is a partial list of the competing demands, drawn from the personal experience of the committee members:

- Focusing on products and markets where there is an experience base *versus* focusing on new technologies or new markets. Focusing on existing products and markets can create a near-term orientation because working on small changes in existing products will pull the time horizons of investment in toward the present.

- Continuing with a project or development direction *versus* pursuing a new project or direction. New and untested ideas can look better than a project that is partially completed, especially one that needs management attention to solve critical problems. Management must resist the temptation to switch capriciously to a new and different project or to stay too long with a project that is failing to bear fruit.

- Incorporating new ideas and information in the execution of a project *versus* holding to an initial direction. Long-term technology development projects are particularly susceptible to "feature creep"; the continuing accumulation of improvements on the basis of new technical or market information without ever freezing a design. The adoption of some new features may be crucial, but the adoption of too many changes will overburden a project with modifications that escalate its cost and delay its market entry so much that it fails to achieve ultimate success. The trade-off is particularly difficult in rapidly growing new businesses based on rapidly changing technology.

- Acting on assessments by involved parties *versus* acting on independent judgment. Project and research directors, responding to company pressures and incentives, routinely underestimate the cost in both time and money to develop new products, produce them, and develop new markets. Underestimating cost and time to payoff on a new technology is a frequent result of combining proponent optimism and real-world uncertainty. Senior managers must establish systems for carefully selecting technology development projects and accurately evaluating their potential benefits, their projected resource needs, amd their realistic schedules (and the uncertainties in forecasts of all three).

Because of the special characteristics of ventures or plans that depend heavily on the development or deployment of new technology, senior management needs to understand technology or to seek out advisers (both inside and, frequently, outside the company) who can offer advice and information with respect to technology development and deployment. The consequences of poor management of a longer-term technology-based strategy appear in a host of discernible events—promising projects that are late to market because they are deprived of resources at a crucial time, an increase in the reliance on older products to produce revenues for the company, a difficulty in attracting and keeping good talent because "the action" is elsewhere, a steady decline in investments in longer-term product or production technology development or longer-lived assets, or, the ultimate indicators, steady declines in market share and the exit of patient shareholders.

Unfortunately, the only formula that top management can pursue to ensure success is the continuous exercise of judgment based on analysis and experience; successful management explicitly plans for the long-term future of a company and manages tensions in execution effectively. Managers intent on, or charged with, developing and deploying technologically new products or processes have to deal with additional uncertainties. In particular, they need to develop and use the ability to assess realistically the uncertainties and time constants of developing technologically new products, processes, and businesses.

Design and Implementation of Career Development Systems and Compensation Schemes

In a nutshell, career systems and compensation schemes may establish procedures or incentives that heavily weight the attention of managers and technical personnel toward the short-term results of segments of the organization rather than on the longer-term results of the company (or business unit) as a whole. Alternatively, such systems and schemes can (1) help focus the attention of management and employees at all levels on the long-term health of the company and the need to make investments in the long-term health and growth of the company; and (2) be an integral part of a company's personnel development efforts—by its very nature a long-term investment.

One crucial relationship between corporate time horizons and careers and compensation exists in the character of job tenure in companies. On the one hand, employees who are promoted too rapidly from one job to another may be encouraged to make overly cautious decisions that may leave their successor with longer-term negative results. On the other hand, they may try to get projects done fast before they move on even if this erodes long-term strength.

Poorly managed turnover can be a substantial barrier to successful technological innovation; turnover, without management attention to the use of deputies or other back-up provisions for key positions, can lead to "start-over." Also, without adequate experience in a position, an individual is not likely to have an appreciation for the fits and starts of commercial technological innovation, a basis for judging the importance, or viability, of a given innovation. On the other hand, significant problems can be created if a company's promotion and career development system leaves managers in certain program or functional positions for too long. Senior management must balance the desire to reward employee performance and to develop talent internally by providing individuals with a variety of experiences against the advantages of longer-term tenure and experience in positions; it is critical to design incentive systems and career paths carefully so that short-term and appropriately long-term performance is rewarded.

Another way in which corporate time horizons are tied to employment practices is through financial incentive systems, which often do not successfully match rewards to employee contributions over the necessary time frame for performance. To ameliorate this problem, bonuses can be tied in part to longer-term performance or be at least in part project-related rather than being based solely on current year corporate (or division, or unit) profitability. Obviously, there is no single correct time horizon of reward systems—incentives need to be tied to each employee's specific assignments and, where appropriate, to important time horizons of the division, the industry, and the particular strategy of the company. On the one hand, many managers' primary responsibility is for day-to-day operations and annual incentive systems based on performance against annual operating plans are most appropriate. On the other hand, if a typical new product development cycle is three years with two more years to payoff, the development team bonuses could be partially based on a five-year assessment, and a division director or senior managers' bonus might reflect the average investment-to-payoff cycle. A short-term "fashion" division might get full cash bonuses based on this year's profits. Longer-time horizons for compensation must, of course, be coupled both to the actual time frame of the work and to the employee's ability to affect the performance of the unit or function on which his or her compensation is to be based. Whenever employees are assigned to specific team efforts, part of their incentive compensation should be tied to the success of the team's efforts.

One strong financial incentive for long-range management thinking is equity ownership. To encourage longer time horizons, however, the value of such equity interest must be related to the long-term health of the company. Stock options, with long expiration times (up to ten years) and long vesting times (up to five years) may be a good way to accomplish this. The more of managers' compensation—particularly that of the senior executives—that is in this form, the more likely they will be to take a long-term view of company operations. The intent is that options should extend well beyond foreseeable departures (retirement or promotion to a different position are two common events) to provide an incentive for long-range decision making.

The focus of changes in incentive systems should be to make the reward system robust. The system should use incentives that include, but are not limited to, financial rewards and should also encourage appropriate long-term performance. The incentive system should, in effect, reward understanding of the technological basis of the industry and decision making that is a result of fundamental understanding of the product and the market (Baker et al., 1988; Baker, 1990; and Jensen and Murphy, 1990). One method of accomplishing this may be to design the incentive system of companies to mimic those of new, independent ventures. Additionally, the incentive system could provide finanical penalties for poor management decisions.

In summary, senior managers—individuals whose responsibility includes long-term development of the company—need to strive for a balance between the movement of personnel for talent development purposes and longer assignments to encourage constancy of purpose and longer-term vision within the company. Additionally, bonuses or incentive pay should be based on performance over a period of time that matches the challenges of the position rather than just on the previous year's performance. This is especially important with regard to business projects linked to a technology development and deployment cycle that can take several years to come to fruition. It is also worth considering incentives to encourage longer tenure by key employees who may work on several different projects over time, for example, bonuses that follow an employee from job to job.

Finally, long-range decision making may be encouraged by providing a substantial portion of compensation in stock options for executives who are in a position to influence the long-range performance of a corporation. To be effective, some portion of such options packages should not be exercisable (or vested) for some period of years to encourage appropriate time horizons (the length of time would depend on characteristics in the specific industry).

Decision-Making Methods and Performance Measurement

The method of selecting projects and making funding decisions can determine the time horizon of investment decisions (Baldwin, 1990; Hertenstein 1988, 1990). Analytical decision-making tools that are exclusively finance based, such as discounted cash flow (DCF) analysis, can, if applied blindly, limit the decision maker's exploration of technology issues. These tools, improperly applied, may lead the decision maker to discount future income streams too much; capital budgeting manuals provide ways to ignore cash streams beyond some time, typically three to five years in the future. Common failures in the application of financial decision tools are (1) to ignore unquantifiable characteristics, such as customer satisfaction; (2) to ignore the impact of unpredictable changes—such as competitor's actions—in preparing the analysis; (3) to pay inadequate attention to the assumptions behind the analysis; and (4) to ignore the second- and third-order factors affecting the consequences of a project or program being considered.

If financial analysis tools are poorly understood or sloppily applied, they can lead an analyst to ask the wrong questions. In the worst case, such tools—because they depend on reducing uncertain events to quantified financial outcomes—can create the illusion that a situation is thoroughly understood simply because it can be modeled. DCF and other similar tools, if misapplied, misused, based on faulty assumptions, or used with incorrect data, can result in funding only low-risk, near-term projects.

A common problem in the application of DCF with regard to technology-dependent projects is the tendency to look only at the cash flows represented by the project itself, without considering what will happen if the project is not undertaken; the value of technological improvements is often based on small net increases in volume or percentage margins relative to current results. On the other hand, technology investments may *prevent* a loss in market share. The cascading consequences of *not* pursuing the technological improvement often include the following:

- Loss of market share
- Lower production volumes
- Slower learning curve progress
- Substantially higher relative costs

The combination of these events can overwhelm the direct effects of the improvement. Conversely, pursuit of improperly evaluated programs based on too optimistic DCF projections have all too frequently led to unnecessary and possibly disastrous dissipation of corporate resources. At a minimum, DCF must give full consideration to the net effects of the investment in relation to the baseline event of not pursuing the project at all.

Technology investments in new projects in many companies are given credit only for the direct effects these projects have on sales volume and profitability. In reality, of course, the shareholder will benefit by some multiple—defined by a stronger price-to-earnings ratio (P/E) of the company—if the project is successful in either an operational *or* a strategic sense. When looking at acquisitions, these P/E potentials are always considered in the analysis, but they are not part of most internal technology development decisions. Consequently, internal technological developments will tend to be discriminated against in relation to acquisitions of companies that might provide new product lines or product extensions. Comparison of the P/E value of an internal project—no matter how difficult or tenuous the estimation—against the baseline case of not performing the project could change the seeming attractiveness of many longer-term technological projects.

Effective use of financial analysis tools can help establish more appropriate time horizons for projects by estimating the future value of technology investments. However, no analytical tool can solve what is in many cases an organizational problem. Analytical tools such as DCF are only as good as the estimates that are developed as input, often provided by biased parties. Careful analysis of the assumptions behind estimates of expenditures and revenues is an important responsibility of management at all levels. In a similar vein, management is responsible for setting the context for good analysis and decision making; casually selected and promulgated hurdle rates, or target internal rates of return, which are not sensitive to the characteristics of a changing economic environment or of the type of project being considered will create bad decisions. In particular, artificially high hurdle rates will preclude necessary (and appropriate) investment in technology or capital equipment. Finally, developing methods to articulate the benefits and the risks of developing new capabilities, with their inevitable uncertainties, need to be considered in sensitivity analyses to make full use of such methods.

The issue of measuring corporate performance poses similar problems, though they relate more to assessments by senior management and the board of directors than to decision making within a company. The choice of measures of performance may seem innocuous, but the choice (or choices) drive a variety of decisions and actions taken by members of an organization. Particularly troublesome is the way in which an accepted measure generates actions intended specifically to support the improvement of that measure. For example, top management may choose to focus its attention on profit margin. While profit margins are useful information in some situations, selection of profit margin as an overriding measure of company performance can promote damaging game playing. In most operations it is possible to maintain—or increase—the profit margin by minimizing expenditures for advertising, market development, R&D, or customer-site product testing (or by "creative" inventory valuations, by holding inadequate reserves and by creating windfalls from financial asset manipulation or the sale of assets). Such actions will unquestionably raise the near-term profit margin but may result in limiting the organization's ability to maintain market share and remain profitable in the long run.

In contrast, the selection of market share as the primary measure of organizational performance drives very different actions. Concentrating on market share may encourage decisions to increase capacity, invest in targeting of the product in the market, undertake extensive product development and improvement, and pursue different pricing strategies than those in the profit-as-goal example. These decisions focus on developing and maintaining long-term staying power in the market.

Executives should consider carefully the measures by which they evaluate the performance of the organization. Simple measurements or metrics of performance, by their very nature, can mask what is actually happening; measuring one characteristic of successful competition does not ensure that there is a causal connection between the success of the enterprise and improvement in that metric.

In summary, managers need to guard against the misuse of investment decision-making or modeling tools based on financial indicators. These tools must be examined and revised continuously to ensure that hurdle rates appropriately reflect the company's cost of capital and risk, that appropriate-horizon values are used, and that appropriate termination values are used. Overly narrow analyses of technology-intensive projects or programs are a particular danger; the focus of project and program analyses should be on evaluation of the effect of entire programs on the performance of the corporation, not on individual or incremental projects. Effective use of project analysis tools applied to technology investments requires that the decision to do a particular project or program be compared with a base scenario of not doing the project or program. The marginal impact of an investment decision should be explicitly compared with the marginal impact of doing nothing. Most important, the analyses should not be restricted solely to financial data—financial analyses should always be supplemented by the use of non-financial analyses.

Finally, senior managers should consider carefully the measures selected for examining and reviewing corporate performance. Executives must recognize that the measures used have a significant impact on the decisions that are made and the actions that are taken. In choosing, for example, quarterly profitability instead of market share as a measure of performance, senior management may be promulgating a set of criteria for performance that is not in the long-run interest of the company or the shareholders. Particularly problematic are those measures of performance that do not support the development and maintenance of long-term organizational capabilities in an industry where such capabilities are a requirement for continuing success.

SUMMARY AND RECOMMENDATIONS

The committee concludes that management practices and corporate governance practices and structures must be regarded as important determinants of a company's investment time horizons and its ability to develop and deploy new commercial technologies. This conclusion is supported simply by consideration of the large number of factors under the control of management that affect company time horizons and the degree to which these factors are not well understood or managed in many companies.

The time horizons a company exhibits are a reflection of corporate goals, directions, and strategies—issues that are the responsibility of corporate governance structures. Boards and the top management they select are the people who consistently have access to the type of information that facilitates a high-level, long-term view; they are uniquely responsible for the company's future. Thus, if a public company's performance is weak because of short-sighted investment behavior, it is a failure of its board of directors.

The board of directors importantly influences the time horizon of a company through the selection and development of senior management. To be effective at governing a corporation, a board of directors should (1) be attentive to the importance of balance within the senior management team and (2) link compensation packages for senior executives to their performance in developing and implementing plans for the long-term prosperity of the company.

The committee recommends that a greater portion of the compensation of managers who are in a position to influence the long-range technological performance of a corporation be granted in stock or stock options. The options should not be exercisable for several years—perhaps five years—and should last for a number of years—perhaps ten years.

Second, since the actions of boards of directors are crucial to the time horizons of a company, so are the methods by which directors are selected, compensated, and removed. Corporate governance might be improved by increasing the financial stake that their outside directors have in the corporation by requiring that they own shares at least equal in value to a specified multiple of their annual fees as directors. The intent is to link board member compensation as directly as possible to long-term performance in stock price, at the same time recognizing that such measures will not be a cure-all. Also, because of the special characteristics of ventures or plans that depend heavily on the development or deployment of new technology, board members' understanding the processes of commercial technological innovation is crucial to the execution of their responsibilities.

The committee recommends that corporate boards have nominating committees operating independently of the CEO in choosing new board members and that these nominating committees, in technology-driven companies, give more weight to technological skill as well as business experience in selecting new board members.

The committee recommends that corporations move to increase the financial stake that their directors have in the corporation and that a signficant part of director's compensation be paid in stock or stock options.

Third, senior company management plays a very important role with regard to time horizons in at least three ways: (1) constancy of purpose coupled with flexibility in the development and execution of corporate strategy; (2) design and implementation of career development systems and compensation schemes that promote attention to longer-term corporate goals; and (3) design and choice of decision-making methods and measurement tools that suit the demands and uncertainties of technology-dependent investments.

The committee recommends that bonuses paid to managers with scope and authority over long-term performance be based not just on the previous year's performance, but on multiple years' accomplishments.

The committee recommends that companies actively reconsider the way they use investment decision-making tools such as discounted cash flow analysis, especially with regard to decisions involving new or continuing investments in technology development and deployment. Faulty or unrecognized implicit assumptions, lack of attention to strategic considerations, and poor handling of technological or market uncertainty in the use these tools can critically damage a company's decision making about technology investments.

4

Time Horizons and Cost of Capital

America's financial climate is not conducive to long-term investments in technology and equipment, compared with Japan, Germany, and the most rapidly developing Asian nations. Several things contribute to this relatively unfriendly environment. High U.S. capital costs shorten the time horizons of investors, so do the pressures exerted on companies by the stock market, particularly by institutional investors and takeover specialists. In sum, both government policies and business practices reinforce an excessive concern with short-term profit in America.

—U.S. Congress, Office of Technology Assessment, Making Things Better: Competing in Manufacturing, *(Washington, D.C., U.S. Government Printing Office, 1990), p. 9.*

There are two related, yet clearly distinct, sets of questions about capital for technology investments. *The first set is linked to the perspective of a single nonfinancial corporation and its interaction with financial markets.*

The committee is indebted to Joseph Morone and Albert Paulson for their excellent work interviewing corporate executives and preparing their report to the committee, "Cost of Capital—The Managerial Perspective," which is published as Appendix A of this report. Morone and Paulson's work helped shape the committee's deliberations with regard to the different perceptions of the importance of the cost of capital in different companies and industries, the ways in which a company can manage its cost of capital, and the importance of a technical and marketplace lead in allowing long-term thinking. As a matter of policy, however, it is important to note that findings, conclusions and recommendations of that report are Mr. Morone's and Mr. Paulson's and are not intended to reflect opinions or judgments of the committee or the National Academy of Engineering.

From the perspective of an individual corporation, competitive performance (market share or profitability, for example) derives in part from successful development and deployment of commercially viable products and services. That depends on investments—investments in things ranging from product research and development through production equipment upgrades and personnel training to market creation or development. Investment capital is a market commodity, and companies seeking to "purchase" capital face a typical set of buyers' options—some of the available products are relatively cheap and some are relatively expensive, some products can come with desirable features and some are packaged with a number of unwanted extras, some products are offered by a reputable dealer and some can be obtained from the financial equivalent of a guy selling watches from inside his trench coat. In other words, as purchasers of capital, companies face a range of choices, most of which affect a company's options for investing the resources in company operations.

As was described in Chapter 2 and discussed in Chapter 3, companies have considerable control over, and latitude in, the way they make investment decisions. In addition, companies affect their internal investment options by a range of interactions with financial markets. The cost of funds, the pretax cost of capital, and internally established investment hurdle rates interact to affect the attractiveness of investments. The important questions, from the company perspective, revolve around whether the cost of capital is a high-priority concern, ways in which corporate actions increase or decrease the company's cost of capital, and noncost consequences (exposure to takeover, for example) of corporate financial decisions.

The second set of questions relates to capital in macroeconomic terms. Technological advance and productivity growth depend on the aggregate amount and efficiency of investments in capital formation, research and development, and human resources. Perhaps the most problematic concern about capital costs from the macroeconomic perspective centers on "invisible" losses to the national economy, reflecting investments not made. At a macroeconomic level, the aggregate national rate of economic growth depends on rates of investment in plant and equipment, in the development of human resources, and in the development and application of technological advance. High national capital costs (relative either to capital costs in other nations or to different times in the same nation) will dampen virtually all investment in assets promising future returns and lead to slower rates of national economic growth. This impact may or may not show up in the performance of individual firms—depending on the industry, competitive position of the firm, and the ability of a company to cope with high relative costs of capital, any individual company may show few ill effects of high capital costs. It is, however, a serious national concern.

Notwithstanding the considerable power of economic theory and empirical work, the extreme complexity and interdependence of this system make it difficult to identify changes in fiscal policy, monetary policy, financial market regulations, or institutional structure that will unquestionably improve the allocation of resources. The task is even harder if the goal is to determine what actions will truly lead to stronger national performance in commercial technological advance. The policy arguments relating technological advance and financial markets have centered on ways to lengthen the time frame of investments by providing lower taxes on investments held for longer periods (e.g., lowering capital gains taxes on investments held for long periods of time); reducing the cost of capital by reducing government borrowing (i.e., deficit reduction); or increasing savings (e.g., shifting toward consumption taxes rather than income taxes). Of overall concern is the continuing fiscal deficit, which places pressure on capital availability in the United States and thereby increases capital costs to the detriment of all investments, long-term and short-term.

These two perspectives—the corporate perspective and the macroeconomic perspective—are linked through financial markets. In general, the organization and functioning of financial markets reflect economic opportunities. Since there is money to be made making microchips, trading wheat futures, or renting apartments, financial instruments and institutions have evolved to allow investors to buy shares of microchip companies, participate in wheat futures trading, or own shares of real estate partnerships. Having said that, it is important to recognize that financial markets do not mirror economic opportunity perfectly or without constraint.

Significant allocation problems arise when competing companies face different expectations on the part of lenders or shareholders (different costs of capital), which many argue has been the case in competition between U.S. and foreign competitors for at least the last 15 years. Tax structures or regulatory policies that unintentionally introduce a bias in favor of investments that pay back quickly can exacerbate the problems introduced by differing costs of capital. Also, information problems abound in financial markets in spite of regular government intervention (Securities and Exchange Commission regulation) designed to protect investors from fraud and market manipulation.

Another issue is that financial markets are as susceptible to structural problems as any other market. Problems arise if financial markets are not organized to collect and deploy capital effectively, an argument made in recent years about the influence of institutional investors in the United States; financial organizations themselves become players, bringing with them all of the decision-making biases and limited rationality of any organization.

In summary, the cost and availability of capital for all investment, as well as the economic efficiency of marginal investment decisions, need to be viewed through two lenses: (1) corporate financial structure and behavior; and (2) economic conditions that depend, in part, on government tax and fiscal policy. Both are inextricably and recognizably linked to financial market structure and the behavior of financial market actors. For example, in a recent survey of 139 members of the Industrial Research Institute, respondents named "general management practices" and "external financial pressures" as the primary causes of "erosion in U.S. technology leadership" (National Science Board Committee on Industrial Support for R&D, 1991). In other words, Wall Street is commonly blamed, in part, for U.S. short-term behavior; nonfinancial corporation executives tend to blame financial intermediaries like banks and institutional investors for increasing short-term pressures on organizations.

The following sections focus on the ways in which financial markets and companies interact to affect time horizons, and they suggest strategies, primarily from the corporate perspective, for improving levels of investment in long-horizon technology development and long-lived productive assets. A short section is included on national differences in costs of capital. Although this study does not take an international comparative approach in other matters it addresses, the committee judged the issue of international capital cost differentials to be to so much a part of the current debate over the role of time horizons in competitiveness that it deserved attention. The section is also useful in that it develops several explanations about the ways in which economic and financial market parameters affect company time horizons.

FINANCIAL MARKETS AND TIME HORIZONS IN THE 1990s

The amount of money controlled by institutional investors has grown significantly over the past two decades. The pool of institutional assets has increased from $569 billion in 1970, to $1,773 billion in 1980, and to $5,810 billion in 1989. These equity assets have grown as a percentage of the New York Stock Exchange (NYSE) from 27 percent in 1970 to 54 percent in 1989 (Salomon Brothers Inc., 1990). At the same time, turnover of securities has dramatically increased. On the NYSE the ratio of volume of shares traded annually to total shares listed has grown from 12 percent in the early 1960s to more than 50 percent in the mid-1980s (Chandler, 1990). This trend—the emergence of a liquid institutional market for corporate control—has been called the "commoditization" of corporate ownership (Jacobs, 1991) and is recognized as an important change affecting the pace and character of restructuring and redirecting U.S. industrial enterprises.

As institutional investment managers hold and manage larger and larger portfolios, they have come to be a substantially more important and influential part of the financial structure of the nation. One manifestation of the growth of an institutionalized market for corporate control is a dramatic change in the character and pace of corporate restructuring. In the 1990s even a large public company can be bought, divided, reconfigured, and sold by individuals or institutions with no previous experience with, or substantial connection to, the company. The leveraged buy-outs, mergers, and acquisitions that were so much a part of the financial news in the 1980s are a product, in part, of the growth of an institutionalized market for corporate control. The highly visible corporate restructurings of the 1980s have created a widespread perception of financially driven corporate restructuring as excessive and wasteful; many people in industry believe the kind of wholesale financial restructuring that took place in the 1980s is the curse of American capitalism. Others, not surprisingly, regard it as a natural and important part of free market competition.

A second manifestation of the increase in institutional investor holdings is the impact that institutional trading has on the behavior of managers in publicly traded companies. Because institutional investors trade increasingly larger blocks of stock and can do this at an increasing rate, these investor's executives are driven to examine the performance of organizations constantly and to make decisions based on that performance. This may force investors' time horizons to be as short as one day. In turn, the argument goes, this can compel executives in publicly traded companies to have shorter and shorter time horizons.[4]

The effects that the rise of institutional investors and the commoditization of corporate control have on the time horizons of corporations are ambiguous. Pressure on corporate managers by institutional investors is, indeed, likely to increase (McCartney, 1990), perhaps creating new pressures on corporate governance; however, it is far from clear whether there is a unbreakable link between institutional investor trading and short-term behavior on the part of companies. For example, as institutions are becoming some of the largest shareholders in individual companies, some are behaving like long-term investors and flexing their muscle to bring about changes in corporate management. The California Public Employees' Retirement System (CALPERS) is perhaps the prime example of an increasingly activist institution, an institution that has pursued the traditional role of a large shareholder in corporate governance.

In general, institutions have been most aggressive with stockholder resolutions, commonly pursuing a social agenda—divestiture of investments in South African companies, environmental practices, and reducing foreign

[4]It is worth noting that institutional investors with short time horizons are often created by the same corporate managers who feel they suffer from the short-term orientation of financial markets. The orientation of some institutional investors toward quarterly performance is, in part, a reflection of the short-term orientation of boards of directors and executive managers of nonfinancial companies who want maximum return at all times in their own pension funds and often insist on measuring corporate pension fund managers on a year-to-year basis. This creates a situation that, in the long term, pressures fund managers to perform to match or beat market index numbers on a quarterly basis. To allow institutional investors to seek longer-run returns, both corporations and governments, in their administration of employee pension funds, can adopt policies that allow them to reward their fund managers for long-term performance rather than on the basis of the quarterly, annual, or, at most, two-year evaluations commonly in use today.

oil-dependency. Recently, however, they have been moving closer to a traditional large shareholder's role. CALPERS garnered attention for a role it played in suggesting and supporting directors for a distressed company and, most recently, for refusing to reelect a board of directors as a protest against large compensation packages for senior management.

Is including institutional investors in the governance of an organization better or worse than relegating their influence to the proxy mechanisms? Is a rapid and organized response from shareholders, if they perceive that management is failing to perform well, a problem or a solution? The nature and type of relationships between institutional investors and companies may be more at fault for "shortsighted behavior" than the incapacity or unwillingness of institutional investors to invest for the longer run.

With regard to the impacts of corporate restructuring, companies with poor short-term financial performance are, it seems, increasingly exposed to the threat of hostile takeovers; in public companies a low stock price based on poor short-term performance can open up a takeover opportunity as the assets of the company can be deployed or sold at higher value than the value of the stock. In general, takeovers and acquisitions have been (must be) financed with debt, loaned against the value of the business. The result is that management is pushed to short-term actions to maintain or maximize cash flow to pay down the debt. To the extent this takes place in a company dependent on investments in expensive and uncertain technology development or deployment (R&D or process investments) it can harm the company's ability to compete effectively by shortening the company's time horizon dramatically. What is unclear is whether such outcomes are bad either for the company or for an economy; a company that has a low stock price because it is overinvesting in foolish or misguided R&D can be helped or mercifully dismembered by a takeover that forces attention to short-term cash return to investors.

Was the frenzy of restructuring in the 1980s good or bad for the long-term performance or technological competitiveness of U.S. firms? The data are either unavailable or inconclusive or both (Coffee, et al., 1988; Flamm, 1990; Ravenscraft and Scherer, 1987). In particular, (1) the story is still unfolding with regard to the large bulk of those companies that took on a heavy debt load; and (2) the majority of takeovers—indeed the majority of all mergers and acquisitions have been in mature industries with little explicit research and development (Grundfest, 1990). What is clear, it seems, is that the amount of corporate restructuring during the 1980s—both "good" deals and "bad" deals—was enormous. Disagreement (or ambivalence) among the financial and business community about the value and impact of commoditization of corporate control is widespread.

In spite of enormously important changes in financial markets in recent decades, individual companies continue to have substantial latitude to affect how much they pay for capital. Companies determine their image in financial markets (and thereby the value of their stock) by their profitability and through choices about issuing new stock, paying dividends, borrowing funds, and retaining earnings. Management of debt-to-equity ratios and other choices about sources and uses of capital are importantly dependent on a corporation's board and senior management. It is the responsibility of the directors, working with senior management, to match capitalization assessment and needs to the corporate strategy. All of these actions affect the company's market cost of capital, leading to the conclusion that companies are themselves important players in determining their cost of capital.

A COMPANY'S CONTROL OVER ITS COST OF CAPITAL

The board of directors and management of public companies are partially responsible for a corporation's cost of capital through their impact on the performance of a company. Management's goal, regardless of strategy, is to create value for the stockholder either through appreciation in the value of equity holdings or in dividends paid to shareholders. The value of the stock determines whether the company will have access to capital, fixes the price of the capital that the firm raises, and influences the dividends that must be paid or the growth of retained earnings needed to hold onto the stockholders. The access to capital provides a firm with the opportunity to invest in projects that are likely to yield returns and affect future performance.[5]

This simple description—although accurate—hides the degree to which a company's prospects (and therefore its cost of, and access to, funds) depend on the characteristics of the company and the business the company

[5]This section deals almost exclusively with public companies. Large private companies have clear advantages in terms of loyal "shareholders" (discussed in the next section), but they face a different set of issues when it comes to raising or using external capital. Smaller private companies face an entirely different set of constraints. As a general rule, small private companies are financed by personal savings, family and friends' investments, bank loans, finance company loans, or individual venture investors; public equity markets play almost no role in financing small companies. Additionally, organized venture capital funds are rarely interested in companies that are not expected to grow rapidly to a size and degree of profitability and promise at which it makes financial sense either to "take" a company public or to sell a major equity stake in the company to a larger public corporation. Although many of the central points of this section (e.g., the importance of managements' skill in communicating effectively with sources of finance) apply to smaller private companies, there is a substantial set of issues, which the committee did not choose to address, that are worthy of further exploration. Do the informal financial markets that serve technologically oriented start-ups in the United States function efficiently? Do small private companies have adequate access to credit for technological modernization? How do the typical sources of finance for these activities affect the time horizons of operations? What is the effectiveness of government programs, such as Small Business Innovation Research, aimed at supporting technological start-ups?

is in, and on such uncertain factors as the rate of overall economic growth, the rate of growth in a particular market, and the performance or likely performance of competitors. The growth potential and degree of risk or uncertainty, as perceived by markets and investors, are major factors that control the availability of capital to firms. One common measure to manage is the price-to-earnings ratio of public securities; the P/E ratio can be kept high through revenue and profitability growth, and the expectation of that growth is one way to maintain long-term stockholders. At an industry level, if the potential for growth and profitability is questionable, the money available to firms in that industry will dry up. The cost of borrowing money to invest in technology development—if it can be obtained at all—will be very high. Conversely, a high growth potential is directly linked to low capital cost.

Financial Markets, Technology, and Company Valuation

Although there is much that managers cannot control about a company's interaction with financial markets, there is an important question over which they have considerable influence: How do investors (markets) know and judge the timing and magnitude of a company's "prospects" when there is significant technological or technology-related market uncertainty? Information about prospective returns creates financial markets, and governments have long been in the business of regulating basic financial information to certain standards of reliability—in the United States, individual states took the lead in such regulation, and the Securities and Exchange Commission was established in 1934. Standard financial information about a company, however, does not begin to provide the information necessary to judge the likely impact of a new long-term corporate research program or a systematic (and probably expensive) in-house effort to bring new production technologies into existing facilities. As such, information becomes a significant problem in the relationships between providers and users of capital. While providers of capital to technology-based enterprises want predictability and the maximum possible assurance of success as well as high returns, the users of capital often need abundant resources and considerable latitude and time to solve technical, organizational, and market problems. It is, by design, an uneasy relationship of mutual interest, lubricated primarily by information.

At one extreme is the venture capital community, which often invests substantially in single-product, technology-based start-up firms, usually with the goal of growing the company to a sufficient size to take it public through an initial public offering of stock. The high risk of such investments—from the investor's perspective—is offset by a variety of structural mechanisms. The most important of such mechanisms is the direct involvement of the venture capitalists in governance of the organization, giving them both the best available information about likely outcomes and significant control over operations (Sahlman, 1990).

At the other extreme is the information exchange between large, multiproduct corporations and the providers of funds, such as banks and stockholders. In some cases, information about technological matters in large companies is easy to interpret—a court judgment granting a disputed patent right or the announcement of a new major R&D or capital investment program—but most often the value of technology-related company actions is clouded by questions of execution (how effective will the company be in turning technology into profits) or lost in the small impact that any single technological development will have on a large, multiproduct company. Quality programs, which for the most part consist of management actions and worker practices in combination with some small design, production, or product technology changes, are a model for long-time-frame, hard-to-interpret, technology-related actions that can substantially affect a company's performance.

The executives in a firm do have some influence over who purchases the stock through effective communication of company information, by management, to money sources. In addition, management can affect how markets perceive the firm's potential as an investment opportunity by establishing long-term relationships with key participants. Most financial market actors—both providers and users of capital—seem to be insufficiently engaged in assessing, analyzing, and valuing technologically uncertain company actions. From the company perspective, strategies for investor relations can more readily allow a company to invest in technological efforts that have a long time frame. From the investor perspective, there may be strategies for information gathering and investment selection that favor long-term, high-return technological investments (Fisher, 1992). Both sets of strategies have the potential to allow a better match between investor preferences and the demands of technical innovation, and both deserve substantial exploration and development by the finance and technology management research communities.

Time Horizons, Technology Investments, and Ownership Structures

Additionally, the ownership structure of the firm may have a direct impact on the cost of capital to the firm; if the firm is private or a large block of the stock is held by a single investor, a family, or a trust, the firm can be provided with more stability than a firm whose stock is virtually all traded openly. Family-owned companies have the reputation, at least in the first generation, of being able to make long-term investments—the investors are willing to be more patient than a firm whose stock is traded in large blocks by institutional investors. Therefore, another aspect of managing the cost of capital arises from the advantages of loyal shareholders. Although most corporations cannot adopt a wealthy family as a long-horizon patron, they can seek to increase the loyalty of their shareholders by the following means, among others:

- Employee stock ownership plans (ESOPs), which can create a block of votes representing the interests of employees (a voting block that is usually assumed to support long-term survival and growth of the company)
- Cooperative R&D arrangements or joint ventures with another company that is particularly skilled in a technology that is important to a venture can reduce the risk of the venture allowing both companies to invest with longer time horizons
- Alliances or partnerships that may have the effect of lowering the cost of capital (teaming up with a firm that has lower cost capital available)
- Cultivation of financial investors who have both a reputation for being, and the expressed intent to become, stable, long-term investors

Cultivation of longer-term investors—institutional or others—requires constant attention and effective use of information transfer. This cultivation also demands constant contact; it is critical that this information flow be maintained in both good times and bad. Too many companies limit their information sharing with stockholders when there is bad news and, consequently, investors are left with an increased level of discomfort about the investment and will eventually sell their shares.

In most cases, managers have many opportunities to affect both the stockholder profile and the investment community's perception of the organization with the effect of lowering a company's cost of capital and thereby providing management with the opportunity to lengthen investment time horizons. In large part, the incentive for senior managers to pursue these strategies depends heavily on their own time horizon. Senior corporate managers should develop and cultivate the following relationships with the financial community and with stockholders:

- Long-term relationships with banks and insurance companies and constant communications with these institutions
- Long-term relationships with analysts, investment bankers, and institutional investors, focusing on constant communications, especially in bad times
- Good relationships with stockholders, providing them with realistic analysis of each of the corporation's major businesses and providing them with warnings of downturns well in advance.

In all of these relationships, managers must manage the conflict between the desire for full disclosure of information and the needs of proprietary secrecy.

THE ECONOMIC COST OF CAPITAL: NATIONAL DIFFERENCES AS A CRUCIAL ISSUE

Numerous studies have shown that, over the past two decades, the market cost of capital in Japan and Germany has been as little as half that in the United States, and that this difference has been and continues to be a source of competitive advantage for companies based in those countries.[6] The lack of low-cost capital in the United States, particularly to smaller companies, relative to that in other countries is argued to create a severe competitive disadvantage for U.S. firms in terms of investment in plant, equipment, and R&D. The difference is often offered as an explanation for the shortsighted behavior of U.S. executives when compared with their foreign counterparts.

Changing world financial conditions—in particular, the increasing liberalization of national capital markets (and concomitant globalization of financial markets, including floating currency exchange rates)—are chipping away at the measurable differences in the costs of debt. In particular, there have been substantial increases in debt flows across national borders but substantially less equity holding than debt holding across national borders. In 1990, for example, gross purchases of U.S. securities by foreigners were $2,120 billion. Of these, $1,947 billion, or 92 percent, were for debt instruments and $173 billion, or 8 percent, for corporate equities (U.S. Treasury Bulletin, June 1991). While debt markets have truly globalized, equity markets continue to have a substantially national character, probably primarily as a result of limited information flows about equity opportunities in other countries. As a result, market clearing rates for debt are nearly equal in the United States and major industrial competitor countries (Hatsopoulos, 1991). This has focused increasing attention on the cost of equity and the impact of corporate financial structures and national financial market structures (both of which vary considerably among nations) on the cost of capital a company faces.

It is clear that the relationships between firms and financial institutions are different in the United States, Europe, and Japan. There are important international differences in the banking system and in the relationships between banks and firms. One often cited difference between the United States and Japan is the role that Japanese banks play in holding equity in a company. In particular, Japanese banks are allowed to hold shares in corporations, allowing them to be both lenders to, and shareholders of, a corporation. In Japan, financial institutions (banks and insurance companies) have held as

[6]See, as examples, Abuaf and Carmody (1990); Cordes (1991); Hatsopoulos et al. (1988); Hatsopoulos and Brooks (1986); and McCauley and Zimmer (1989). It is important to note that the most common result of these studies—that U.S. capital costs are substantially higher than capital costs in Japan—has been questioned in a number of studies. See, as examples, Kester and Luehrman, 1989; and Nachbar, 1990.

much as 40 percent of total outstanding corporate shares (Kester, 1986; Federal Reserve Bank of San Francisco, 1991). In Germany, also, banks may play a substantially more important role in the governance of nonfinancial corporations through supervisory boards involved in the day-to-day affairs of borrowers.

The capital and ownership structure of many Japanese corporations reduces the risk of both lending funds and holding equity and serves to lower the real cost of capital to firms. In financial terms, the nation's financial structure is such that risk of holding equity in a Japanese corporation is lower than the risk of holding equity in a seemingly comparable company (size, asset base, competitive prospects) in the United States. Referring to the earlier discussion of risk and return (the capital market line), it is clear that mechanisms that reduce financial risk also reduce the rate of return expected by investors and, hence, the cost of equity capital for the corporation. Particularly noticeable is the degree to which risk-reducing financial arrangements involving banks, companies, and the government (Kester, 1986)—arrangements that would be abnormal and perhaps illegal in the United States—allow some Japanese firms to operate with high relative debt/equity ratios in their financial structure and still maintain stability in operations.

For a variety of reasons, many U.S. firms borrowed heavily in the 1980s and substantially substituted debt for equity in their corporate financial structures (Blair, 1990). According to Benjamin M. Friedman (1990), "On average during the 1950s and 1960s, it took 16 cents of every dollar of pre-tax (and pre-interest) earnings to pay [U.S.] corporations' interest bills. The corresponding average for the 1970s was 33 cents. Since 1980 it has been 56 cents. In no year since 1981 has the interest share of earnings been below 50 cents on the dollar." Highly visible private leveraged buy-outs and threats of takeovers were particularly evident signs of this trend in the 1980s. This rush to leverage during the 1980s moved many U.S. firms toward debt/equity positions similar to Japanese firms but without the benefits of a financial system organized to minimize the risk of operating in such a manner. What is clear is that high debt/equity ratios (high leverage) in U.S. firms can substantially increase a firm's susceptibility to business cycles (Cantor, 1990). Highly leveraged companies tend to be more unstable, having greater cyclicality in their investment and employment. This includes instability in capital investment and in R&D investment and more and higher cycles of hiring and firing employees. Clearly, in this situation a company would be extremely unlikely to have a long-term perspective; time horizons would be driven to the very short term.

It is beyond the committee's scope to settle the various uncertainties about the existence and magnitude of national differences in costs of capital or the impact of different types of corporate financial structure and restructuring on long-term investment. Most of the uncertainties are likely to continue until more and better data exist and researchers better understand the various situations and trends. There are, however, emerging areas of apparent agreement.

In particular, it appears, on the one hand, that national differences in market rates for debt are not likely to exist except for relatively short-lived fluctuations arising from national economic or monetary policies. On the other hand, however, national differences in the cost of equity are likely to persist at some level. Differences in national equity costs will be sustained where they are introduced (1) as an intended or unintended effect of differences in corporate capital and ownership structures and practices; or (2) as an intended or unintended effect of national policies that isolate national financial markets or that direct and subsidize investment; or (3) where significant information asymmetries among national equity markets exist.

This does not present an optimistic scenario for U.S. firms with regard to capital cost differentials. If the structural relationships among Japanese banks and companies (or European banks and companies) continue to produce stable but more highly leveraged capital structures in Japan and Europe than in the United States, then for the foreseeable future—even with all trade barriers removed and the cost of debt equalized—the "playing field" will remain inherently uneven. Japanese and European firms will continue to have a competitive advantage of investing with longer time horizons and potential access to more patient capital than their U.S. counterparts.

A variety of U.S. government policies have the potential to reduce the cost of equity capital in the United States:

- *Reduce the federal deficit.* Federal budget deficits are a burden on financial markets and, as such, drive up real interest rates. Some combination of government spending cuts and new tax revenues are required to reduce the federal deficit. The direct impact of reduced federal budget deficits would be reductions in the cost of debt, with indirect impacts on equity costs.
- *Reduce or alter capital gains taxes.* Taxes on returns on investments drive a wedge between the return the market demands for a particular level of risk and the pretax return an investment must generate (i.e., an investment must return an amount that pays both the tax and the investor). Some industrialized countries currently tax capital gains at a much lower rate than in United States; a lower rate of U.S. capital gains taxation (or a rate that decreases substantially as the investment is held longer) would reduce the pretax return demanded, reduce the cost of equity capital, and lengthen time horizons. There are a host of well-developed schemes for changing capital gains taxation (prospective, retrospective, indexing, etc.) each of which has its advantages and disadvantages (Congressional Budget Office, 1991a; Hatsopoulos, 1989; Shoven, 1990).

- *Eliminate or reduce "double taxation" of corporate profits.* Corporate profits are taxed first as corporate profits and then as individual income (dividends or appreciation at sale). Double taxation of corporate profits has been fully or partially eliminated by the United Kingdom, France, Germany, Canada, Japan, and Australia through "dividend relief" schemes or by the existence or establishment of low capital gains tax rates. Reduction in this double tax wedge between before- and after-tax rates of return would lower the cost of equity capital in the United States. Proposals to eliminate double taxation or to mitigate its impact on time horizons include (a) taxing income generated from equity sales on a graded scale such that the tax rate decreases with increases in holding time; and (b) providing to sellers of equity a tax credit per share sold that is equal to the U.S. federal income tax paid by the corporation on that share during the period of the seller's ownership of the share.
- *Remove barriers to risk-reducing financial arrangements.* Some of the risk-reducing financial arrangements prevalent in other countries (allowing bank ownership of corporate stock, for example) or aspects of government regulation that affect corporate structures (antitrust laws, for example) may be amenable to change without endangering the health of the financial system or exposing the public to monopolists.
- *Promote household savings.* At a fundamental level, a relatively high national cost of equity is a reflection of a relative preference by citizens for consumption rather than savings. Tax policies such as those allowing all wage earners to make before-tax contributions to Individual Retirement Accounts (instituted in 1981 and eliminated in 1986) have been offered as ways in which to shift the preferences of consumers toward savings with an expected increase in the availability of capital for long-term investment.

Such actions are sensible and necessary if the United States is to remain an attractive place to produce for global markets and to avoid disadvantaging existing U.S. firms in global competition. However, such moves are unlikely to *eliminate* remaining differences in the capital costs faced by U.S. and foreign competitors. Rather, policies that reduce the cost of capital in the United States may be able to remove as much of the difference in relative costs of capital as is possible given the different financial market structures and financing arrangements in different nations. U.S. companies should (1) take advantage of opportunities to reduce any relative equity capital cost differential by tapping global financial markets; and (2) prepare themselves to operate with a cost disadvantage in the area of equity capital.

There is a wide range of options by which to attempt to lower equity capital costs in the United States, and the NAE study committee chooses not to endorse particular proposals; it is beyond the scope of the committee's expertise to evaluate the likely impact of alternative policies on important aspects of national priorities such as the distribution of the tax burden, the impact on federal government revenues, and the stability of U.S. financial markets and institutions. Such matters should be carefully, explicitly, and promptly weighed and acted upon by the executive and legislative branches of government.

SUMMARY AND RECOMMENDATIONS

The cost and availability of capital for all investment, as well as the economic efficiency of marginal investment decisions, can be viewed through two lenses: (1) corporate financial structure and behavior; and (2) economic conditions that depend, in part, on government tax and fiscal policy. Both are inextricably and recognizably linked to financial market structure and the behavior of financial market actors.

The emergence of a liquid institutional market for corporate control, or the "commoditization" of corporate ownership, is an important change affecting the pace and character of restructuring and redirecting U.S. industrial enterprises. As institutional investment managers hold and manage larger and larger portfolios, they have come to be a substantially more important and influential part of the financial structure of the nation. The growth of an institutionalized market for corporate control has driven dramatic change in the character and pace of corporate restructuring and is changing the relationships among boards of directors, top corporate management, and institutional investors.

Even in the context of these strong trends in capital markets, corporate senior managers and boards of directors continue to be able to influence both the structure of company ownership and the way in which markets perceive the firm's potential as an investment opportunity. Most financial market actors—both providers and users of capital—are weak at assessing, analyzing, and valuing technologically uncertain company actions. From the company perspective, strategies for investor relations can more readily allow a company to invest in long time-frame technological efforts. From the investor perspective, there may be strategies for information gathering and investment selection that favor long-term, high-return technological investments.

Additionally, the ownership structure of the firm may have a direct impact on the cost of capital to the firm; if the firm is private or a large block of the stock is held by a single investor, a family, or a trust, the firm can be provided with more stability than a firm whose stock is virtually all traded openly. Corporations can seek to increase the loyalty of their shareholders, and therefore the time horizons of decisions, by such activities as employee stock ownership plans, and cultivation of financial investors who have a reputation for (and the expressed intent of) being stable, long-term investors.

In most cases, managers have many opportunities to affect both the stockholder profile and the investment community's perception of the organization and thereby lower a company's cost of capital and provide management with the opportunity to lengthen investment time horizons. In large part, the incentive for senior managers to pursue these strategies depends heavily on their own time horizon.

The committee recommends that managements and boards of directors of companies dependent on long-horizon technological developments (a) implement investor-relations strategies that aggressively and clearly communicate the technological prospects of a company; (b) work to develop long-term relationships with lenders and equity investors; and (c) aggressively pursue joint ventures or other arrangements to reduce the risk of specific technological ventures.

With regard to international differences in the cost of capital, national differences in market rates for debt are not likely to exist except for relatively short-lived fluctuations arising from national economic or monetary policies. However, national differences in the cost of equity are likely to persist at some level.

The committee recommends that the federal government move to allow longer investment time horizons for U.S. corporations through tax policy changes designed to reduce the pretax cost of equity capital.

A variety of policy actions have been proposed to lower the relative cost of capital in the United States, but such moves are unlikely to *eliminate* remaining differences in the capital costs faced by U.S. and foreign competitors. U.S. companies should (1) take advantage of opportunities to reduce any relative equity capital cost differential by tapping global financial markets; and (2) prepare themselves to operate with a cost disadvantage in the area of equity capital.

5

The Influences of Government Investments and Regulatory Policies on Corporate Time Horizons

The United States is actually neither as innocent of nor as unskilled at industrial policy as many Americans seem to believe. In his "Report on Manufactures" of 1791, Alexander Hamilton gave classical expression to what is today a commonplace of industrial policy theory: the understanding that market prices are important and effective signals for adjusting supply and demand in the short run but that they are quite inadequate as guides for investment decisions about new technologies, choice of products, and scales of production ten to fifteen years hence. Hamilton wrote, "Capital is wayward and timid in lending itself to new undertakings, and the State ought to excite the confidence of capitalists, who are ever cautious and sagacious, by aiding them overcome the obstacles that lie in the way of all experiments."

—Chalmers Johnson, "Introduction: The Idea of Industrial Policy," in C. Johnson, ed., The Industrial Policy Debate *(San Francisco, ICS Press, 1984), p. 17.*

Government polices and investments are a pervasive, important, and often positive influence on the business environment and economic development of any industrialized nation. The following are among the many government polices and actions affecting the business environment:

- The structure of taxes (several aspects of which were discussed in chapter 4)
- The design and implementation of workplace and environmental regulations

- The amount and nature of government support for generic technology development, research, and programs too large for single firms or with payoffs too far in the future or too uncertain to attract private capital
- The amount and nature of government investments in physical infrastructure and human capital
- The legal environment of operating a business encompassing, among other issues, the protection of intellectual property rights and the handling of liability claims

Through these and other roles, government plays an important, varied, often obvious but sometimes subtle part in determining the time horizons of corporate investment decisions. The impact of government polices and actions on business investment in technology and operating practices is the subject of a vast and continually growing body of scholarly literature and policy studies (see, as recent examples, Carnegie Commission on Science, Technology and Government, 1991; Council on Competitiveness, 1991; Lee and Reid, 1991; and Porter, 1990). A comprehensive review of the literature and current debates in even one or two of these areas—environmental regulation or product liability, for example—could easily run to several hundred pages and would require expertise not represented on the current study committee; a comprehensive treatment of the influences of government on corporate time horizons is clearly beyond the scope of the committee's work. Therefore, recognizing the diversity, complexity, and importance of these issues, and aware of the limitations of time and expertise, the committee has chosen to focus on two types of government influence on corporate investment horizons, neither of which is widely understood. First is the role of government in providing a stable environment for investment, including the role the government plays in the creation of markets. Second is the role of government in investing in complementary public assets—national, regional, or local public assets, which work in tandem with private investment to allow and drive economic growth.

GOVERNMENT POLICIES AND A STABLE ENVIRONMENT FOR INVESTMENT

As discussed in Chapter 2, too much uncertainty is the natural enemy of long-term investment. Frequent upheavals in the marketplace or uncertainty about the terms and directions of competition add a significant element of risk to longer-term business decisions, which drives companies to seek recovery of their investments in the shorter period of time and dampens investment in activities that, by their very nature, will take substantial time to come to fruition. Federal, state, and local governments play a crucial role in the affairs of industry. The policies, routines, and practices of governments can either improve or erode predictability (decrease or increase risk) in markets and technologies and thereby determine whether an environment is conducive or inimical to long-term investment.

The Mixed Impact of Regulation and the Legal Environment

On the one hand, stable and predictable regulation for worker and consumer safety and protection of the environment can drive important and innovative developments with positive long-term consequences. Waste and emissions standards establish fixed targets for improving processes and, as such, can encourage innovative approaches to problem solutions; for example, product innovation in the automobile industry to reduce pollution has resulted in major innovations. Providing incentives to minimize wastes in industrial processes not only may improve the environment but also may reduce production costs significantly. The result of these regulations can be to create a reliable "playing field" for competition, thereby improving the long-term health and capability of these industries, and to increase competitiveness in foreign markets. On the other hand, frequent changes in tax policy, regulatory structures, government licensing practices, and other forms of government interaction with industry can be quite damaging.

Various types of, and approaches to, regulatory and legal structures directly and indirectly affect the time horizons of organizations in different ways. Licensing procedures, patent lives, work place safety regulations, and environmental regulations can either extend or constrain the time horizons of organizations depending on the situation and the manner in which government laws and regulations are implemented.

Product liability concerns, for example, are often cited as a legal constraint that can indefinitely lengthen the payback time for new product development projects by creating significant uncertainty about a company's ability to recover investments. When this happens the increased risk to an investment increases a company's cost of capital. A legal system that inhibits longer-term investments because of its unpredictability, delays, and punitive treatment of product liability issues will hamper economic growth. It is impossible to foresee all circumstances in which a new product will be used; technological and economic advances must depend to some degree on "caveat emptor." On the other hand, an effective product-liability system can offer customers redress against genuinely fraudulent or unsafe products that make it to the marketplace—a safety net that will make customers more likely to trust producers' explicit or implicit claims and therefore more quickly create a predictable market for a product. In short, effective tort law is a balancing act, which, depending on its implementation, can lengthen or shorten corporate time horizons.

Time horizons are also strongly affected by the protection of intellectual property rights, both domestically and across national boundaries. For example, if inventions are denied effective patent or trademark protection, investment is likely to flow toward nearer-term product (or service) modifications rather than toward R&D investment for new (unprotectable) innovations. The diligence and effectiveness of the government in the protection of intellectual property rights—both domestically through the U.S. legal system and internationally through trade negotiations and other international treaties—can have a significant impact on long-term planning and R&D investment on the part of corporations in a number of industries.

Among the government policies and actions that are the most consistently damaging to long time horizons are those which create disincentives for long-term planning and investment. Late or uncertain promulgation of environmental and workplace standards often unnecessarily diverts company investment capital from longer-term technology development. Similarly, the inability of companies to plan on predictable and rapid resolution of licensing, plant siting, or environmental or health and safety clearances creates greater uncertainty for companies, delays returns on investments, and decreases company's willingness to take longer-term investment risks. Slow or inefficient government regulatory processes discourage otherwise productive investments by the private sector.

The Government's Role in the Creation of Stable Markets

Among the ways in which governments promote long-term investment is the role they play in the creation of markets or marketplaces. First, the government's considerable buying power has created predictable markets for "public" goods, some of which have become private goods. Commercial passenger and freight aircraft, created in part by government investments in, and demand for, defense aircraft, are a classic example. Additionally, markets for private-sector weather prediction and monitoring, environmental monitoring and waste disposal, public health systems, or large-scale satellite, computer, or networking systems are based on, or were supported by, markets created by government purchases, often in combination with government R&D.

Second, the use of regulation to create or stabilize markets is an important public role in encouraging long-term investment. Government regulation plays an important role in creating safe and reliable financial and air transport markets, albeit the definition of safety in the two markets is quite different. Government's ability to create a monopoly (often regulated and designed to be temporary) during certain stages of an industry's development is another tool to promote long-term investment. This tool has been used with AT&T and the U.S. telephone system as well as with innumerable local activities such as electric power, gas, water, sewer, and taxicab services (which are often regulated through commissions) or commercial real estate development (controlled through local zoning laws).

Third, the government plays a crucial role in the creation of stable markets through its role in setting formal or de facto standards. As new markets and technologies emerge and develop, standards are often unclear or in constant flux. At some point—when necessary standards and potential technologies become clear—government helps establish formal standards, or participates in setting de facto standards, by becoming a buyer and thereby promoting long-term investments in the developing industry. Such interventions must be carefully timed to avoid freezing the system too soon or too late, but they can be enormous successes.

In summary, the government-created regulatory and legal environment has a substantial impact on time horizons of companies, but the impact is complex and multidimensional; some regulations and legal procedures can lengthen corporate time horizons, while other regulations, or legal constraints that introduce substantial unpredictability, can shorten time horizons. Government procurement and regulatory policies have clearly provided initial and sometimes large markets for a host of products that we think of as "commercial" today. Early, persistent, and effective participation by government in many markets has given the United States a host of industries built on long-term investment that would have developed more slowly, if at all, without government participation. The importance of government policies with regard to the regulation and creation of markets needs to be acknowledged, and expertise in the use of such policies to support long-term investment should be cultivated.

The government should make sufficient investments in its own expertise and in evaluation and improvement of systems to reduce significantly the time spent in carrying out such fundamental governmental responsibilities as environmental approval of new facilities, obtaining licenses on government controlled or regulated technologies, obtaining patent approvals, getting product approvals through the Food and Drug Administration, and obtaining final determinations on technology-based court cases. The government should consider substantial investments in court or arbitration infrastructures specialized in these matters to shorten the time cycles for resolution of product and process liability, environmental impact, and other regulatory cases. The intent of such investments would be to encourage efficiency and timeliness in the prosecution of government regulatory and legal processes.

Finally, it is important to recognize the degree to which changes in policy that affect business are driven by events beyond the control of legislators or executive branch policymakers; the process of making government policy is, by its very nature, subject to many fits and starts, uncertainty, and changes in direction. Although there is no single or simple change that can (or should) alter the nature of policymaking in democratic governments, stability in policies has considerable intrinsic value and the disturbance of a stable business environment should be regarded as a cost (sometimes small but often large) of any change in policy that directly affects business investment decisions.

GOVERNMENT INVESTMENTS, COMPLEMENTARY ASSETS, AND PRIVATE-SECTOR TIME HORIZONS FOR INNOVATION

David J. Teece, a professor of business administration at the University of California, Berkeley, has used the term *complementary assets* to describe the variety of capabilities or assets that support an innovation:

> In almost all cases, the successful commercialization of an innovation [technical knowledge about how to do something better] requires that the know-how in question be used in conjunction with other capabilities or assets. Services such as marketing, competitive manufacturing, and after-sales support are almost always needed. These services are often obtained from complementary assets that are specialized. For example, the commercialization of a new drug is likely to require the dissemination of information over a specialized information channel. In some cases, as when the innovation is systemic, the complementary assets may be other parts of the system. For instance, computer hardware typically requires specialized software, both for the operating system and for applications. Even when an innovation is autonomous, as with plug-compatible components, certain complementary capabilities or assets will be needed for successful commercialization (Teece, 1987, pp. 70-71).

The concept of complementary assets is particularly useful in understanding the role of government investments in private-sector time horizons; complementary assets are the publicly provided infrastructures or services that permit, support, or work in conjunction with private investments in physical or human capital or R&D. Public infrastructure and publicly supported research and development are two important examples.

Infrastructure

Traditional, physical, public infrastructure systems are the most well recognized form of complementary assets. For example, government land grants, bond guarantees, and regulations designed to develop the nation's transportation infrastructures—starting with canals and roads and then railroads—created important, stable, complementary assets that allowed and supported the development of agricultural, manufacturing, and retail businesses. As the era of modern transportation and communications began, the government provided eminent domain for telegraph and telephone communications and helped coordinate standards for wireless radio, and later satellite communications. It subsidized and regulated the development of airports, air routes, and Federal Aviation Administration communications and control systems that make today's air linkages possible. Federal and state governments also made enormous investments in the land-grant university systems that became the core intellectual resources for the United States during its period of agricultural expansion and industrialization.

In more recent years, governments have been the principal investors in roads, dams, waterway maintenance, hospitals, schools, public health, outdoor recreation, space, and defense systems. The agricultural and mass-distributed foods industries in the United States have been helped significantly by heavily supported government agricultural research, land use, water development and agricultural extension services, standards for foods and packaging, and enforced systems of standard weights and measures. The government's provision of infrastructures, implicit subsidies, and direct markets have provided the longer lead times, risk capital, and stable markets for a wide range of industries.

Projects like the Tennessee Valley Authority, the Rural Electrification Administration, and the Interstate Highway System not only created the original jobs and profits from these projects. Each opened up huge new markets that otherwise could not have been reached by product producers (such as radio, television, appliance, automobile, trucking, manufacturers). Relatively small initial investments (subsidized by government) opened up whole economic regions to be major markets and producers for modern business and home technologies, creating huge economic multiplier effects for the whole country. Such investments also led to U.S. primacy in the kinds of construction these projects represented, the products they allowed to be produced, and the services infrastructures they fostered. In macroeconomic terms it is well documented that the ratio of capital invested to gross national product is a key ingredient in both economic growth and competitiveness. Such generalizations extend to both the public and the private sector. In recent years, the ratio of the total federal government budget dedicated to investment has fallen steadily in relation to transfer payments and services entitlements. Specifically, the inflation-adjusted amount of government spending on physical infrastructure, about $26.2 billion in 1990, is about equal to its 1980 level (Congressional Budget Office, 1991b). It is important to note that level, inflation-adjusted expenditures over long periods do not describe a state of consistent levels of federal support; as GNP grows a constant level of spending will represent a smaller investment relative to the demands of the economy.

Research and Development

Government investments in risky or long-term research are the basis of another set of complementary assets that the government provides for businesses: access, at little or no cost, to scientific and engineering information and resources paid for by government. It is well documented that, in many cases, federal government research and development have established much longer time horizons for technological development than individual industries or companies might have been able to exhibit.[7] Such public investments reduce the risk of related private investments and affect an enormous variety of industries, both directly and indirectly:

- The pharmaceutical and medical products industries have been strongly supported by many years of basic research through the National Institutes of Health and other agencies (supported by a strong product patent system and the demands created by government-supported health care).
- The extensive support the federal government has provided since the early 1960s in microbiological, genetic, plant, environmental, and human health research is now beginning to produce a biotechnology industry and the insights that will transform medical care, agriculture, and many industrial and waste-disposal processes.
- Through its long-term research on materials and propulsion technologies, plus the provision of large-scale testing facilities, the government created assets of crucial value to the U.S. aircraft industry.
- The government's early investment in large-scale computers and information networks for atomic and missile research provided the groundwork for today's computer infrastructure, which has given many educational institutions and research units a significant competitive advantage over their counterparts in most other countries.
- By allowing AT&T Bell Laboratories (before the dismemberment decision) semimonopoly privileges and the right to collect a user fee from telephone customers and to use this fee in advanced communications research, the government helped create very long-term investment time horizons in communications, and for years those long time horizons gave the United States a world leadership position in this technology.
- The Defense Department's continued drive to find the highest possible performance materials and systems for military purposes has pushed ahead the frontiers of today's microengineering, test equipment, fiber-polymer composites, and scanning tunneling microscopy imaging capabilities.

---

[7] The literature on the role of government R&D in economic development is vast and continues to grow. Much of the literature is empirical, describing the characteristics, success, and failures of different government R&D support activities. See, for example, Brooks, 1986; Ekelman, 1988; Ergas, 1987; Flamm, 1987; Gelijns and Halm, 1991; National Research Council, 1987; Nelson, 1982; and Rothwell and Zegveld, 1981. The record of government in funding the longer-term research and development that produce information or resources of value to private firms is well documented, though the analytical description offered usually relates to risk rather than time.

- The Agriculture Department's long-term investments in agricultural research led to many of the hybrid seeds, plants, and agricultural techniques that individual or corporate farmers could never have developed themselves.
- Government's coinvestments in satellite systems for weather predictions, communications, and navigation made such systems possible long before they would have been strictly "economic" from a private investor's viewpoint.

Long-term investments in basic research and large-scale systems—and the government's tax and patent encouragement for such investments in the private sector—have proved to have long-term payoffs. The government can also lengthen technology investment horizons by such actions as coinvesting in consortium arrangements for post-basic, but precompetitive, generic technologies (such as materials research, micro- or nano-manufacturing, as well as special engineering and manufacturing equipment that cannot provide an attractive commercial return on investment), developing data bases on medical care outcomes, or supporting experimental mass processes for waste treatment and disposal.[8] Private industry often will not tolerate the combination of low probability, long time to payoff, and high risk or ambiguity of commercial success that these investments require. This is, in part, because the sponsoring company cannot capture the full benefits of a successful result even if it is achieved, since it can, at best, only share the market. However, since society as a whole does capture such full benefits, it is often rational for governments to support such activities when private enterprise could not.

Performing research and development in universities creates further complementary assets beyond the research benefits themselves. These assets stem from the upgrading of university faculty, the advanced training of students, and the diffusion of knowledge that results from publication and from students later building on their personal knowledge base from the projects.

In summary, the government has the scale and stability of revenues to support the development of a wide variety of complementary assets, assets that allow private companies to adopt longer time horizons for their investment decisions, and without which many important industries would not have developed to their current degree. Successful investments in certain technological areas can open multiple secondary and tertiary industries and markets (as government investment has done in such areas as rural electrification, communications systems, semiconductors, and hybrid crops). Other

---

[8]Generic technologies are also known as leverage technologies, the primary characteristic being their applicability in the production of several, or many, different end products or services.

investments, particularly those in infrastructures such as roads, transportation systems, or water systems, create relatively predictable long-term returns for the society by lowering specific costs or increasing productivity and flexibility in forecasted forms.

Time Horizons and the Public-Sector Investment Portfolio

Although it is not typically regarded in such a manner, it is possible to conceive of substantial portions of the federal budget as a national public goods capital budget or an investment portfolio. Although transfer payments and expenditures for current consumption (personnel expenses for the military, for example) make up a the majority of the federal budget, a substantial portion of the budget is devoted to investments in long-lived public assets, such as physical infrastructure or research and development. In some cases, such investments create important public goods that have few direct uses for citizens and companies beyond their stated purpose (e.g., building a military base creates "national security" with only spin-off economic effects). Other types of public investments, however, can be regarded as creating public goods that are also substantial complementary assets for private enterprises.

History shows that the federal government's investment portfolio has allowed or driven the development of important new technologies and, with the support of state and local governments, has funded physical infrastructures that could not have been justified on a return-on-investment basis by any single company or industry. Additionally, the federal government and, to a lesser extent, state and local governments, have provided funding and management for projects that were too large and had completion times too long for any single corporation or consortium of private enterprise. By investments in these and future technological areas, the government can influence the time horizons, development, and competitiveness of U.S. industries in the future by assuming some of the risk of developing new technologies and providing risk capital.

Another crucial long-term opportunity for government investment is education—an investment that upgrades the intellectual capacity of the society and the flexibility of its human resources. The payoffs from investments in education accrue over long periods; they are captured by the individual or society over at least the full lifetimes of those who receive the education. Such investments have a significant impact on overall U.S. productivity, competitiveness, and quality of life.

As mentioned earlier, current federal spending for physical infrastructures is about equal to its 1980 level. Federal outlays in 1990 for education, training, and employment and social services are about 20 percent below an all-time high level of $52 billion in 1979. Research and development spending,

$67 billion in 1990, has considerably increased since 1980, with relatively steady increases in nondefense spending over that time period. Although it is extremely difficult to get a clear picture of federal government spending for long-lived assets, a very rough estimate is that about 20 percent of the federal budget represents investments in education and training, R&D, construction, or other long-lived assets.[9]

These investment portfolio decisions are active subjects of debate by a large, diverse, and knowledgeable set of interested parties. However, the debate and decision-making process could be improved by ongoing evaluation by the legislative and executive branches of the degree to which federal budgets, as proposed and approved, include investments that truly provide physical or human capital, or a knowledge base, for the future. In addition to the intrinsic value of such investments, they are a crucial government contribution to lengthening the time horizons of private-sector investment decisions.

SUMMARY AND RECOMMENDATIONS

Government polices and investments are a pervasive, important, and often positive influence on the business environment and economic development of the United States. The focus of this chapter is the role that government investments and regulatory policies play in determining private company investment time horizons.

With regard to regulation and the legal framework for business, government policies can play a crucial role in creating a stable environment for investment. Frequent upheavals in the marketplace or uncertainty about the terms and directions of competition add a significant element of risk to longer-term business decisions, a condition that drives companies to seek recovery of their investments in the shorter period of time and dampens investment in activities that, by their very nature, will take substantial time to come to fruition. Federal, state, and local governments play a crucial role in the affairs of industry, and the policies, routines, and practices of governments can either improve or erode predictability in markets and thereby determine whether an environment is conducive or inimical to long-term investment and business growth.

The government-created regulatory and legal environment has a substantial impact on time horizons of companies, but the impact is complex

[9]All data in this discussion are from *How Federal Spending for Infrastructure and Other Public Investments Affects the Economy,* Congressional Budget Office, Congress of the United States, July 1991. The estimate of the portion of the federal budget that goes toward long-lived assets is based on an effort at totaling the budget accounts in which at least 50 percent of the budget appears to be dedicated to developing or purchasing long-lived assets.

and multidimensional; some regulations and legal procedures lengthen corporate time horizons, while other regulations, or legal constraints that introduce substantial unpredictability, can shorten time horizons. Government regulation and procurement policies have clearly provided initial and sometimes large markets for a host of products that we think of as "commercial" today. The importance of government policies with regard to the regulation and creation of markets needs to be acknowledged, and expertise in the use of such policies to support long-term investment should be cultivated.

The committee recommends that the federal government invest in improving the efficiency and timeliness of its regulatory, patent, and licensing procedures.

The government should consider substantial investments in court or arbitration infrastructures specialized in these matters to shorten the time cycles for resolution of product and process liability, environmental impact, and other regulatory cases. The intent of such investments would be to encourage efficiency and timeliness in the prosecution of government regulatory and legal processes.

With regard to government investments, the government creates complementary assets—publicly provided infrastructures or services that permit, support, or work in conjunction with private investments in physical or human capital or R&D. Such assets can reduce the risk of related private investments and allow private companies to adopt longer time horizons for their investment decisions. Publicly supported research and development and public infrastructure are two primary examples.

The federal government's investment portfolio has allowed or driven the development of important new technologies and, with the support of state and local governments, has funded physical infrastructures that could not have been justified on a return-on-investment basis by any single company or industry. In addition to the intrinsic value of such investments, they are a crucial government contribution to lengthening the time horizons of private-sector investment decisions.

The committee recommends that the budgetary process for the federal government include more explicit consideration of the degree to which federal expenditures support the creation of long-lived physical and human capital or a knowledge base. Preference should be given to those expenditures that will generate returns for long periods of time and contribute to lengthening the time horizons of private-sector investments in the development and deployment of technology.

6

Summary Argument: Understanding Time Horizons and Technology Investments

It is clear that a large and diverse set of factors affects the time horizons exhibited by U.S. companies, and different factors clearly impinge more or less forcefully, depending on the industry segment and the size and structure of the individual company under consideration. The findings of this study can to be regarded as dividing the sources of short-term behavior into three types: those affecting the *economic ability* of companies to invest, those affecting the *willingness* of boards of directors and managements to use the resources in that manner, and those affecting the company's *operational performance* in effectively investing in technological development and deployment. In other words, some factors prohibit companies from taking a long-term view, others affect the relative attractiveness of long-term investments, and still others—many internal to a company—affect the ability of a company to succeed at long-term programs and projects.

With regard to *economic ability*, there is a valid concern that the nation will underinvest in industries or technologies that take a long time to develop or have long production lead times. Particularly troublesome is the degree to which the relatively higher cost of equity capital results in invisible losses to the nation—investments in long-horizon projects that are not seriously considered. National differences in the cost of debt, which received much attention from those concerned about international competitiveness during the early 1980s, seem to have narrowed. However, differences in the cost of equity across national boundaries are likely to persist, creating a durable but hopefully not crippling disadvantage for those U.S. companies competing with companies that draw their equity capital from other markets.

Government has several important roles affecting the economic ability of companies to invest in long-horizon technology developments. Most obvious, of course, are tax and fiscal policies that can either improve or erode the ability of companies to invest in technological development and deployment. In addition, however, it is important to recognize that government regulatory processes, legal processes, and investments in assets complementary to privately held assets can either improve or erode the willingness and ability of companies to invest in long-horizon projects. In many easily identifiable cases, the government—by creating a market, providing risk capital, or supporting the development of new technologies or large, complex systems—has demonstrated its important role in reducing the risk of long-term development of new technologies and industries.

Despite government actions to improve the economic ability of companies, however, a substantial challenge will remain for most companies. A company's financial structure and financial practices affect its economic ability to invest for the long term, and corporations that are highly sensitive to capital costs need to work to minimize either the existence of equity cost differentials or the impact of such differentials on corporate competitive abilities. Also, the international competitive success of many U.S. companies in long-horizon businesses implies that a cost of capital that is high relative to competitors will not necessarily be an overwhelming disadvantage in company competitiveness. This implies that there is considerable slack to be taken up by better business and technology management practices and increased company productivity; while capital cost disadvantages can have a serious impact on a company's time horizons, it is not clear that the disadvantage is so great that, in many cases, it could not be overcome (or its influence reduced) by better governance and management.

With regard to *willingness* to invest and *operational performance*, the record of U.S. company performance shows that among companies operating in the same capital environment, some are successful at maintaining adequate investments with long time horizons. In other words, while external influences like the cost of capital are important influences on company time horizons, most of the responsibility for success or failure in managing long-horizon investments must fall on the company itself; ultimately, if a company is performing poorly in competition because of near-term orientation in its investments, it is probably because of failures of governance or management or both. The same diagnosis applies if a company is performing poorly in competition because it consistently bungles technological investments—it is a failure of corporate management or corporate governance or both. *Indeed, the two types of failures are intimately related; a company that is ineffectual at developing and applying technology will have short time horizons for technology-related investments.*

Finally, the willingness and ability of public companies to invest in projects with long-term payoff is being affected by sweeping changes afoot in the financial economy of the United States. In particular, the "commoditization" of company ownership appears to have driven or allowed a pattern of corporate financial restructuring—widespread mergers and acquisitions, hostile takeovers, management leveraged buy-outs, and the defenses against such—that *may* be inimical to building a corporation with the ability to invest successfully in technological projects or programs with returns accruing several years out.

It is important to mention explicitly two issues that this report does not address. An important question, not addressed in this analysis, is whether different types of industries—for example, batch production industries; industries in which there is extremely cyclic demand; capital goods industries; commodity industries; service industries; or contracting industries—are subject to different influences affecting the time horizons of decisions. Further, this report deals primarily with larger, public companies. A different set of conditions and influences may impinge on small private companies, whether they are high-tech start-ups or manufacturers in mature industries. Also, although the committee is in agreement about the important directions for government action, the specifics of regulatory reform to improve efficiency or of budgetary analysis to promote investment in long-lived assets clearly need considerably more analysis than is possible in this report.

References

Abuaf, Niso, and Kathleen Carmody. 1990. The Cost of Capital in Japan and the United States: A Tale of Two Markets. Salomon Brothers. Financial Strategy Group. July.

Baker, George P. 1990. Incentive contracts and performance measurement. May 18. Photocopy.

Baker, George P., Michael C. Jensen, and Kevin J. Murphy. 1988. Compensation and incentives: Practice vs. theory. The Journal of Finance 43(3):593-616.

Baldwin, Carliss Y. 1990. How capital budgeting deters innovation—and what companies can do about it. Harvard Business School. July. Photocopy.

Blair, Margaret Mendenhall. 1990. A surprising culprit behind the rush to leverage. The Brookings Review (Winter):19-26.

Brooks, Harvey. 1986. National science policy and technological innovation. Pp. 119-167 in The Positive Sum Strategy, R. Landau and N. Rosenberg, eds. Washington, D.C.: National Academy Press.

Cantor, Richard. 1990. Effects of leverage on corporate investment and hiring decisions. Federal Reserve Bank of New York Quarterly Review (Summer 1990). Photocopy.

Carnegie Commission on Science, Technology, and Government. 1991. Technology and Economic Performance: Organizing the Executive Branch for a Stronger National Technology Base. September.

Chandler, Alfred D., Jr. 1990. Scale and Scope: The Dynamics of Industrial Capitalism. Boston, Mass: Belknap Press of Harvard University Press.

Coffee, John C., Jr., Louis Lowenstein, Susan Rose-Ackerman, eds. 1988. Knights, Raiders, and Targets: The Impact of the Hostile Takeover. New York: Oxford University Press.

Congressional Budget Office, Congress of the United States. 1991a. Capital Gains Taxes in the Short Run. A CBO Study.

Congressional Budget Office, Congress of the United States. 1991b. How Federal Spending for Infrastructure and Other Public Investments Affects the Economy. A CBO Study. July.

Cordes, Joseph W. 1991. International differences in the cost of capital: A discussion paper prepared for the Board on Science, Technology and Economic Policy, National Research Council. March.

Council on Competitiveness. 1988. Picking Up the Pace: The Commercial Challenge to American Innovation. Washington, D.C.: Council on Competitiveness.

Council on Competitiveness. 1991. Gaining New Ground: Technology Priorities for American's Future. Washington, D.C.: Council on Competitiveness.

Dertouzos, Michael L., Richard K. Lester, Robert M. Solow, and the MIT Commission on Industrial Productivity. 1989. Made in America: Regaining the Productive Edge. Cambridge, Mass.: The MIT Press.

Ekelman, Karen B., ed. 1988. New Medical Devices: Invention, Development, and Use. National Academy of Engineering/Institute of Medicine. Washington, D.C.: National Academy Press.

Ergas, Henry. 1987. Does technology policy matter? Pp. 191-245 in Technology and Global Industry, B. R. Guile and H. Brooks, eds. Washington, D.C.: National Academy Press.

Federal Reserve Bank of San Francisco. 1991. FRBSB Weekly Letter. March 29.

Fisher, P. A. 1992. The interface between manufacturing executives and Wall Street visitors—Why security analysts ask some of the questions that they do. Pp. 137-148 in Manufacturing Systems: Foundations of World-Class Practice, J. A. Heim and W. D. Compton, eds. Washington, D.C.: National Academy Press.

Flamm, Kenneth. 1990. Appendix A - Industrial research and corporate restructuring: An overview of some issues. Pp. 121-135 in Corporate Restructuring and Industrial Research and Development. Washington, D.C.: National Academy Press.

Flamm, Kenneth. 1987. Targeting the Computer: Government Support and International Competition. Washington, D.C.: The Brookings Institution.

Friedman, Benjamin M. 1990. Implications of increasing corporate indebtedness for monetary policy. National Bureau of Economic Research. NBER Reprint No. 1421.

Gelijns, Annetine C., and Ethan A. Halm, eds. 1991. The Changing Economics of Medical Technology. Medical Innovation at the Crossroads, Volume II. Committee on Technological Innovation in Medicine, Institute of Medicine. Washington, D.C.: National Academy Press.

Grundfest, Joseph. 1990. M&A and R&D: Is corporate restructuring stifling research and development? Pp. 3-20 in Corporate Restructuring and Industrial Research and Development. Washington, D.C.: National Academy Press.

Hatsopoulos, George N. 1989. Capital gains differential: Does it work? Paper presented at conference "Saving - The Challenge for the U.S. Economy" sponsored by the American Council for Capital Formation, Center for Policy Research, Washington, D.C., October 11-13, 1989.

Hatsopoulos, George N. 1991. Technology and the cost of equity capital in technology and economics. Pp. 65-76 in Technology and Economics, Papers commemorating Ralph Landau's service to the National Academy of Engineering. Washington, D.C.: National Academy Press.

Hatsopoulos, George N., and Stephen H. Brooks. 1986. The gap in the cost of capital: Causes, effects, and remedies. Pp. 221-280 in Technology and Economic Policy, R. Landau and D. W. Jorgenson, eds. Cambridge, Mass: Ballinger Publishing Company.

Hatsopoulos, George N., Paul R. Krugman, Lawrence H. Summers. 1988. U.S. competitiveness: Beyond the trade deficit. Science 241:299-307.

Hayes, Robert H., Steven C. Wheelwright, Kim B. Clark. 1988. Dynamic Manufacturing: Creating the Learning Organization. New York: Free Press.

Hertenstein, Julie H. 1988. Bridging the gap between financial analysis and financial value. Harvard University. August. Photocopy.

Hertenstein, Julie H. 1990. Designing procedures to disclose value in capital budgeting: A field study. Presented at the International Conference on Research in Management Control Systems, London Business School, July 10-12, 1989. Working Paper, revised April 1990. Photocopy.

House, Charles H., and Raymond L. Price. 1991. The returning map: Tracking product teams. Harvard Business Review (Jan-Feb):92-100.

Jacobs, Michael T. 1991. Short-Term America: The Cause and Cure of Our Business Myopia. Boston: Harvard Business School Press.

Jensen, Michael C., and Kevin J. Murphy. 1990. CEO incentives—It's not how much you pay, but how. Harvard Business Review (May-Jun):138-153.

Johnson, Chalmers, ed. 1984. The Industrial Policy Debate. San Francisco: ICS Press.

Johnson, Elmer W. 1990. An insider's call for outside direction. Harvard Business Review (Mar-Apr):46-55.

Kester, W. Carl, and Timothy A. Luehrman. 1989. Real interests rates and the cost of capital: A comparison of the United States and Japan. Japan and the World Economy 1:279-301.

Kester, W. Carl, and Timothy A. Luehrman. 1990. The Price of Risk in the United States and Japan. Harvard Business School Working Paper 90-050. October. Also in Japan and the World Economy 3(1991):223-242.

Landau, Ralph, and Dale W. Jorgenson, eds. 1986. Technology and Economic Policy. Cambridge, Mass: Ballinger Publishing Company.

Lee, Thomas E., and Proctor P. Reid, eds. 1991. National Interests in an Age of Global Technology. Committee on Engineering as an International Enterprise, National Academy of Engineering. Washington, D.C.: National Academy Press. 1991.

McCartney, Robert J. April 8, 1990. New takeover tactic: Ballots instead of bonds. Washington Post. H1, H10.

McCauley, Robert N., and Steven A. Zimmer. 1989. Explaining international differences in the cost of capital. Federal Reserve Bank of New York Quarterly Review (Summer):7-28.

Nachbar, John H. 1990. The cost of capital in the United States and Japan: A survey of some recent literature. Rand Corporation Report N-3088-CUSJR. September.

National Research Council. 1987. Agricultural Biotechnology: Strategies for National Competitiveness. Committee on a National Strategy for Biotechnology in Agriculture, Board on Agriculture. Washington, D.C.: National Academy Press.

National Science Board Committee on Industrial Support for R&D. 1991. Why U.S. technology leadership is eroding. Research & Technology Management (Mar-Apr):36-42.

Nelson, Richard R., ed. 1982. Government and Technical Progress. New York: Pergamon Press.

Patton, Arch, and John C. Baker. 1987. Why won't directors rock the boat? Harvard Business Review (Nov-Dec):10-18.

Porter, Michael E. 1990. The Competitive Advantage of Nations. New York: Free Press.

Ravenscraft, David J., and F. M. Scherer. 1987. Mergers, Sell-Offs, and Economic Efficiency. Washington, D.C.: The Brookings Institution.

Rothwell, Roy, and Walter Zegveld. 1981. Industrial Innovation and Public Policy: Preparing for the 1980s and the 1990s. Westport, Conn: Greenwood Press.

Sahlman, William A. 1990. The structure and governance of venture-capital organizations. Journal of Financial Economics 27(2):473-521.

Salomon Brothers Inc. 1990. The Role of Institutional Investors in Corporate Governance. Working Paper.

Shoven, John B. 1990. Alternative tax policies to lower the U.S. cost of capital. Business taxes, capital costs and competitiveness. American Council for Capital Formation Center for Policy Research. July.

Steele, Lowell W., and N. Bruce Hannay. 1985. The Competitive Status of U.S. Industry—An Overview. Washington D.C.: National Academy of Press.

Teece, David J. 1987. Capturing value from technological innovation: Integration, strategic partnering, and licensing decisions. Pp. 65-96 in Technology and Global Industry, B. R. Guile and H. Brooks, eds. Washington, D.C.: National Academy Press.

U.S. Congress, Office of Technology Assessment. 1990. Making Things Better: Competing in Manufacturing. Washington, D.C.: U.S. Government Printing Office.

U.S. Treasury Bulletin. June 1991. U.S. Department of the Treasury, Washington, D.C.

APPENDIX A

Cost of Capital—The Managerial Perspective

JOSEPH MORONE AND ALBERT PAULSON

ABSTRACT

This study, prepared for the National Academy of Engineering's Committee on Time Horizons and Technology Investments, examines whether the cost of capital in this country is viewed by senior executives as a source of competitive disadvantage. In particular, it examines the impact of the U.S. cost of capital on managerial decision making. We interviewed 15 senior executives from four capital-intensive industries. Five of the 15 executives (Group 1) argued that the cost of capital has a major impact on their ability to compete, and directly constrains their investments in capital equipment and R&D. Another eight executives (Group 2) expressed a roughly opposite view, arguing that the cost of capital is not one of the primary competitive factors in their businesses. The remaining two executives (Group 3) took an intermediate position.

The differences in views of the impact of the cost of capital on U.S. competitiveness led us to consider two additional questions: how do the Group 1 firms manage this potential source of disadvantage, and what are the possible reasons for the differences in views between Groups 1 and 2? The interviews reveal a rich array of tactics for coping with what the

---

This study was prepared for the National Academy of Engineering's Committee on Time Horizons and Technology Investments. We are indebted to the committee members, to Kathryn Jackson, NAE fellow and study director, and to Bruce Guile, NAE program director, for the help and support they provided in carrying out the study. We especially wish to thank the executives cited in this study for taking the time to discuss with us their views on the cost of capital. Findings, conclusions, and recommendations of this report do not necessarily reflect opinions or judgments of the committee or the National Academy of Engineering.

A version of this paper appeared in *California Management Review* 33 (Summer 1991):9-32.

Group 1 executives view as the high U.S. cost of capital. And two factors in particular appear to account for the different views: the firms in Group 2 tend to be in better financial condition than those in Group 1, and are technology leaders in their respective markets.

While there is heated dispute about whether or not the cost of capital in the United States is higher than it is in Japan, there is a growing consensus that Japanese firms behave as *if* their cost of capital is lower. Moreover, it appears this behavior is deeply rooted in Japan's industrial structure and financial culture, and therefore, is likely to endure. How can U.S.-based firms remain competitive over the long haul against competitors who are able to behave as if they have a lower cost of capital? This analysis suggests that the answer lies in the actions of both individual firms and the federal government that enhance or preserve U.S. technology leadership.

The time horizons of American management have become a subject of widespread concern. A growing body of evidence suggests that the rate of investment by U.S. industry, although it is higher than it has been in the past, is not keeping pace with that of our leading competitors, particularly the Japanese.[1] The pattern, if true, is cause for alarm. Over the long run, inability to keep pace in rate of investment results in declines in competitiveness and in turn, in standard of living. But while there is growing concern that U.S. industry's time horizons are shorter than Japan's, there remains widespread debate about the reasons for the difference. The most severe point of contention arises over the cost of capital and its relationship to the problem of time horizons. On the one hand, there are those who argue that the cost of capital in this country is higher than it is in Japan, and that this difference lies at the heart of the time horizons problem. The logic of this argument is as follows: since the cost of capital determines the threshold or "hurdle" rate of return required to justify an investment financially, the relatively high cost of capital in this country makes investments that are necessary from a competitive point of view difficult if not impossible to justify financially. Further, these investments are being made by our competitors abroad where the cost of capital is lower. Over time, this difference in rates of investment builds into a significant competitive disadvantage for U.S. firms in capital-intensive industries. There remains widespread debate within this school of thought about exactly how large the cost of capital differential is, and about the reasons for the difference, but there is general agreement that reducing the difference substantially would have a major impact on the competitiveness of U.S. industry (Hatsopoulos and Brooks, 1986; and McCauley and Zimmer, 1986).

A second school of thought, which has not received nearly as much prominence as the first, doubts that there is a persistent difference between the U.S. and Japanese costs of capital, and that the reasons that such differences appear to exist are rooted in difficulties associated with measuring and comparing costs of capital. This view does not deny that there may be important differences between U.S. and Japanese time horizons, but suggests that the explanation for such differences lies less with differences in cost of capital than with differences in the nature of corporate ownership and governance (Abuaf and Carmoy, 1990; and Kester and Luehrman, 1989, 1990b).

This paper is about the cost of capital, but it makes no attempt to resolve the continuing debate about whether or not there is a persistent difference in the U.S. and Japanese costs of capital. Instead, it approaches the subject from a behavioral perspective. We explore not the economic meaning and measurement of the cost of capital, but the way it is perceived by and influences senior executives in a range of American firms. We begin with the following premise: whether or not the cost of capital is lower in Japan than it is in the United States, many Japanese firms appear to approach their investment decision making with longer time horizons than their U.S. counterparts. That is, they behave *as if* they enjoyed a lower cost of capital (Kester and Luehrman, 1990a). This poses obvious difficulties for their U.S. based competitors. The purpose of this paper is to take a first step toward understanding how senior executives in U.S. firms view this problem and deal with it. In particular, this paper addresses the following questions:

- Do executives see the U.S. cost of capital as an important source of competitive disadvantage? Does it constrain their making what they believe to be strategically or competitively important investments?
- If so, how do they balance the conflict between the constraints imposed by the U.S. cost of capital and strategic requirements of their businesses?
- And, if we find differences among executives in their views on the competitive significance of the U.S. cost of capital, how do we account for those differences?

The answers to these questions are of considerable practical concern regardless of which school of thought proves correct with respect to the cost of capital. Either our Japanese competitors enjoy a lower cost of capital than U.S. firms, or they behave as if they enjoyed a lower cost of capital. The underlying causes for this difference—whichever view is correct—appear to be widely and deeply rooted in the industrial structure and financial culture of the two countries, and while the two systems seem to be converging, substantial differences are likely to persist for a long time to come (see examples in Kester, 1991; Zielinski and Holloway, 1991). Even among those who insist that there is a significant differential in the cost of capital, there is a growing recognition that this differential will not be easily eradicated and that the problem is far more complex than simply a matter of equalizing interest rates. The differential, they argue, stems from among other factors: differences in capital structures, differences in relationships among firms and between firms and banks, differences in the tax treatment of capital gains and depreciation, differences in accounting practices, and because of the differences in relationships between banks and firms, substantial divergences between published and actual interest rates.[2]

So whether or not there is a cost of capital difference, the reasons that lead Japanese firms to behave as if they had a lower cost of capital are likely to be with us for a long time to come. Even with trade barriers removed and interest rates equalized, Japanese firms are likely to continue to invest with long time horizons. How to compete effectively against competitors with such long time horizons becomes one of the central issues for U.S. industrial competitiveness.

APPROACH

We examined these questions through interviews with senior executives at 15 firms drawn from four industries. Our criteria for selecting industries were that they be (1) under competitive stress, (2) capital intensive, and (3) relatively diverse. We defined capital intensity broadly, so as to include not just plant- and equipment-intensive industries, but also R&D-intensive and working capital-intensive industries. We attempted to ensure diversity by selecting industries that differed along such broad technology-related dimensions as length of product life cycle, research intensity, and nature of manufacturing process (e.g., chemical process versus component assembly and fabrication). The industries selected in this fashion were machine tools, steel, semiconductors, and pharmaceuticals.

Industry	*Life Cycle*	*R&D Intensity*	*Manufacturing*
Machine Tools	Short	Low	Fabrication/Assembly
Steel	Long	Low	Process
Pharmaceuticals	Long	High	Process
Semiconductors	Short	High	Process

Within each of these industries, we attempted to interview four firms. On the assumption that size might have an important impact on views regarding cost of capital, our goal was to select two relatively large and two relatively small firms for each industry. (Size of a firm was measured relative to other firms in that industry.) Since privately held firms are likely to face very different sorts of financial pressures than publicly held, we limited our sample to the latter. For similar reasons, we excluded firms that had recently been involved in leveraged buyouts, takeovers, or mergers.[3]

In total, 15 (of 16) firms agreed to participate:

	Machine Tools	*Steel*
Large	Cincinnati Milacron Cross & Trecker	Armco Inland Steel Industries
Small	Hurco Companies	Nucor Worthington Industries
	Semiconductors	*Pharmaceuticals*
Large	Intel Texas Instruments	Merck & Co. Pfizer
Small	Analog Devices Chips & Technologies	Amgen Centocor

THE INTERVIEWS

Question 1: Is the Cost of Capital an Important Source of Competitive Disadvantage?

We approached this question by focusing on specific investments that each firm is reported to have made recently or to be planning. For example, Nucor recently invested $270 million in an innovative and controversial continuous thin-slab cast flat-rolled steel-making facility. (As of 1989, book value of Nucor's assets was $1 billion; sales were $1.3 billion.) In our interview with F. Kenneth Iverson (chairman and CEO of Nucor), we discussed how he viewed the impact of the cost of capital on his decision making for this particular investment. As a consistency check, we then attempted to discuss with each executive two related issues—his or her general decision-making style and his or her view of priorities for government policy.

With regard to decision-making style, we attempted to explore the impact and role of discounted cash flow (DCF) analysis with respect to investment decision making. We expect that virtually all firms engage in this kind of analysis in the course of their investment decision making, just as we expect all firms to be driven by the desire to earn, in the aggregate, rates of return that exceed cost of capital-based hurdle rates of return. The question here is, how heavily is their decision making about specific individual investments governed by such financial analysis as opposed to assessments about overall performance? We would expect decision makers who place heavy weight on the results of such analysis in their decision making about specific investments to be more concerned about the competitive impact of the cost of capital than those who describe themselves as balancing the consideration of strict financial analysis with other, more qualitative factors. In other words, decision-making style can serve as a very rough, second measure of executive views on the cost of capital.

In addition, we asked the executives to identify the most important policy issues affecting their businesses to discover more about their perceptions of cost of capital. We expect that firms that view cost of capital as an important source of competitive disadvantage are more likely than their counterparts to include policies related to interest rates and capital formation among their priorities.

The majority of the responses concerning these executives' views of the cost of capital as a source of competitive disadvantage fall into two general groups. Five executives indicated that it is an important handicap, and that the relatively high cost of capital is an important, if not driving, factor in their decision making. Three of these five executives also indicated that DCF analysis weighs heavily in their investment decisions (the fourth and fifth did not discuss decision-making style explicitly), and four cited interest rates and capital formation as their top or among their top policy concerns.

In a second group, another eight executives expressed a roughly opposite view; they argued that the U.S. cost of capital is not a major source of disadvantage and has not been constraining their investment decision making. These executives point to other factors as more important determinants of their ability to remain competitive. All eight suggested that their decision making about specific investments is tempered by considerations other than strict financial projections of rates of return—which should in no way be taken to suggest that they are not driven to earn overall rates of return that exceed hurdle rates—and six specifically explained that the uncertainties associated with their businesses make financial projections about specific investments too unreliable to serve as the driver for decision making about them. None of the eight stressed interest rates as a critical policy issue, although one did argue that the overall economic environment was an important policy concern.

The remaining two CEOs expressed an intermediate position, arguing that the U.S. cost of capital is not a driving factor in their businesses today, but only because they have taken deliberate steps in the past to avoid becoming capital intensive.

Group 1: Cost of Capital is a Critical Competitive Factor

- Inland Steel Industries: To Frank W. Luerssen, chairman and CEO, "managing capital is the name of the game" in the steel industry. Based on his extensive interactions with Inland's Japanese partner (Nippon Steel—

see below), Luerssen is convinced that the U.S. cost of capital is substantially higher than Japan's, and that this puts Inland at a distinct disadvantage. The company has just concluded a 20-year assessment of its business and is setting out on a long-term modernization program. The pace at which it invests and the particular investments it makes or forgoes, are directly constrained by hurdle rates determined by this relatively high cost of capital. Inland must "focus our resources on the sure bets," whereas its Japanese partners, given their distinctly (in Luerssen's view) lower cost of capital, "can justify just about anything that comes along."

- Armco: Robert L. Purdum, chairman, president, and CEO, makes a similar argument. Steel companies operate in a low-growth, low-profitability environment, yet they must continually invest heavily in capital equipment to remain competitive. Compounding the problem is the need to invest in pollution control equipment, which effectively generates negative return on investment. (Purdum does not question the need for such investments, though he does argue for "reasonableness" in environmental policy.) Like Inland, Armco has developed close relations with a Japanese firm (Kawasaki—see below), and like Luerssen, Purdum argues that his Japanese partners operate with far lower costs of capital and therefore are far less constrained in their investment decision making. The higher U.S. cost of capital thus becomes a critical issue, and as with Inland, governs the specific investments that are and are not made. Purdum distinguishes between three sets of investments—those that lead to zero or negative return but are required by government regulation; those that clearly satisfy hurdle rates; and those that are strategically important but that fail to satisfy hurdle rates. It is the third group of investments that are subject to much more review and delay and are pursued only "when we absolutely must," because of the relatively high cost of capital.

- Texas Instruments (TI): TI is in the semiconductor memory business, which is even more capital intensive than steel. Jerry R. Junkins, chairman, president, and CEO, explained that the average investment required for a new dynamic random access memory (DRAM) manufacturing facility is at least $250 million and will grow to $600 million in the relatively near future. The minimum amount of process-related R&D required to remain competitive, let alone lead, is $100-$200 million per year. TI, based on its own internal studies, concluded that the cost of capital of the U.S. semiconductor industry is roughly 75 percent higher than that of its Japanese counterparts. This leads to a clear and direct disadvantage in a business where success is in large measure determined by ability and willingness to invest. The result is that, like his counterparts in the steel industry, Junkins must continually wrestle with a Hobson's choice: make investments that can be financially justified in the lower cost of capital environment of his competitors, but that do not satisfy his own company's hurdle rates; or forgo such investments and pursue what becomes over the long run "a going-out-of-business strategy." The dilemma becomes acute during industry downturns, since experience demonstrates that to remain viable in an environment with low cost of capital competitors (or with competitors who behave as if they had a low cost of capital), the pace of investment must be maintained.

- Cincinnati Milacron: James A. D. Geier, chairman of the Executive Committee, views cost of capital as one of several related factors that have led to the demise of the U.S. machine tool industry. Cincinnati Milacron estimates that the Japanese machine tool industry received roughly $11 billion in subsidies from the Japanese government over a 30-year period. An important component of those subsidies was low-cost capital. Conversely, the higher U.S. cost of capital has deterred investment in plant and equipment. "We would be much more modernized [and therefore more competitive] had the cost been lower."
- Cross & Trecker: The challenge facing Cross & Trecker, at least in the short run, is of a different nature. Its vice chairman and new CEO, Norman Ryker, is restructuring the business and has been focusing his efforts on increasing the company's working capital—a critical issue in a business where orders can take up to a year and a half to complete and where progress payments are more the exception than the rule. To increase working capital, Cross & Trecker recently raised $50 million by issuing preferred stock (10 percent dividends), so Ryker believes he has a very real appreciation of the meaning of high cost of capital. As part of its restructuring, the company is selling some of its businesses and facilities and is not at present investing in new plant and equipment. However, Ryker recognizes that foreign competitors have considerably more modern plants and equipment, and that over the longer term, the company will have to launch a major capital investment campaign. Cost of capital will become an important factor, particularly for "a low-margin industry like machine tools."

As implied in their responses above, three of these five executives expressed a clear orientation toward DCF-type analysis in their decision making. While these analyses do not exclusively govern their decision making, quantitative calculations are emphasized heavily, and appear to be used in these firms to sort out competing investment candidates. This is consistent with the importance that these companies attach to the high cost of capital and their selection of important public policy issues. Luerssen and Junkins cited interest rates and capital formation as their top policy concerns; Purdum cited trade policy, reasonableness of environmental regulations, and policies that would lower the cost of capital, in that order; Geier emphasized overall policy on competitiveness, arguing that policies that influence the cost of capital would follow once the larger problem of competitiveness was confronted; only Ryker did not mention policies that influence capital formation, focusing instead on policies that promote exports.

Group 2: The Cost of Capital Is Not a Major Competitive Factor

Eight other executives offered a strikingly different view of the impact of the U.S. cost of capital on their decision making.

- Analog Devices, Intel: Ray Stata, chairman and president of Analog Devices and Gordon E. Moore, chairman of Intel, do not feel that their decision making—either with respect to R&D spending or capital investment—is seriously affected by the U.S. cost of capital. Moore argued that time horizons in the U.S. microelectronics industry are unquestionably shorter than those of the Japanese. Nonetheless, he does not believe the cost of capital is at the root of the problem. It is a "second-order effect." "There may be some investments [not made by Intel in the past] that we would have made if the cost of capital had been free, but not many." Rather, in businesses "where we have to stay in, almost all our investments are made because they are strategically important." Stata stated that while cost of capital is a factor that needs to be taken into account, it is "at the bottom of the list [of such factors]." Compared to the cost of quality, for example, the cost of capital has a "trivial" impact. Again, investments tend to be driven by strategic requirements.
- Merck & Co., Pfizer: These two pharmaceutical companies do not view the relative cost of capital as an important competitive factor. For Gerald D. Laubach, president of Pfizer, U.S. health policy has considerably more impact on the competitiveness of pharmaceutical companies than any differences in the cost of capital. Indeed, Judy Lewent, vice president for finance and chief financial officer of Merck & Co., and Francis Spiegel, senior vice president of financial and strategic planning, doubt that there is a persistent, significant difference in the cost of capital for U.S. and Japanese pharmaceutical firms. They conclude that the argument that relative cost of capital is a source of competitive disadvantage "is another myth." Rather, both Merck & Co. and Pfizer believe that their competitiveness is driven by the productivity of their R&D efforts. "The name of the game," to use Laubach's phrase, is to pursue development of a portfolio of new drug candidates through the highly uncertain, R&D-intensive, decade-long new product development cycle. It is the productivity of these R&D efforts, far more than any differences in cost of capital, that determine competitive success. This is not to say that Merck & Co. and Pfizer do not pay a good deal of attention to earning aggregate rates of return that exceed their costs of capital. On the contrary, as Spiegel put it, "we are tremendous slaves to the cost of capital. . . . The hallmark of the corporation is to . . . earn cash flow in excess of your cost of capital." This concern appears to be heightened by the relatively high cost of capital in the pharmaceutical industry, which stems from the high levels of risk associated with new drug development. Nonetheless, Lewent and Spiegel see no incompatibility between high costs

of capital and long time horizons. "We don't think strategically in terms of quarters. We think in terms of 5 to 10 years. . . . We don't have much sympathy for people who complain about quarterly pressures. Our mentality and our business drive us to be strategic, but we would be strategic even if we were in a different business."

• Amgen, Centocor: George B. Rathmann, chairman emeritus of Amgen, and Hubert J. P. Schoemaker, chairman and CEO of Centocor, view the high cost of capital from the perspective of biotechnology start-up companies. The problem of generating capital to fund the long, expensive process of developing new drugs has been a driving concern for these executives ever since the inception of their companies nearly a decade ago. When Schoemaker and his associates began their business, they felt they would need $300 million to $500 million in capital to "become a pharmaceutical company." To date, they have raised $350 million. "Finance is the biggest barrier to entry into the pharmaceutical business." Nonetheless, neither executive views the relative cost of this much needed capital as a major competitive factor. As Rathmann described, "when you are in the creation business, it doesn't really matter whether the cost of capital is 4 percent, 9 percent, or 12 percent. . . . You are pushing for such a big payoff that whether the cost is a few percent higher or lower really doesn't make any difference; it's trivial. I don't spend a second's worth of thought worrying about whether we got the capital at a reasonable rate." Indeed, both firms have relied heavily on R&D Limited Partnerships (RDLPs) to fund development of some of their most promising new drug candidates. Investors in RDLPs expect a 40 percent return! As Schoemaker put it, you have to have "a pot of money, or you are not in the game"—whatever the cost of that "pot."[4]

• Nucor and Worthington Industries: Unlike their counterparts at Inland and Armco, neither F. Kenneth Iverson, chairman and CEO of Nucor, nor John H. McConnell, chairman and CEO of Worthington Industries, see the U.S. cost of capital as an important factor in their investment decision making. Iverson argued that aggressive capital spending is a strategic necessity. The company that does not maintain its pace of investment—whatever the differences in cost of capital between U.S. and foreign competitors—falls behind, and once a firm falls behind, it must take on high levels of debt to catch up. In an industry as cyclical as steel, high debt levels eventually lead to serious difficulties. Nucor's rate of investment is constrained not by the cost of capital, but by its overall debt level. It treats a 30 percent debt-to-asset ratio as a ceiling. McConnell likewise emphasized strategic factors in his discussion of Worthington Industries' investment practices. "We try to find the best technology, stay ahead of the competition, and serve the customer. . . . We'll make any investment that will pay back quickly . . . but if it is something that we really see as a must down the road, payback is not going to be that important."

Just as this second group differs from the first group on the competitive importance that they attach to the relative cost of capital, so too they appear to be less driven in their decision making about individual investments by DCF analysis, as is illustrated by the following comments:

- Lewent and Spiegel explained the role of financial analysis in new product development at Merck & Co.: "At the portfolio level [i.e., in the aggregate] we do calculations of return, but it's to make sure we are matching resources to areas where we expect payoffs . . . and to make sure overall return is in line with our expectations for return." But at the individual product level—and development of a successful new product requires on the order of $230 million in R&D, spread over more than a decade[5]—DCF-style analysis does not become a factor until development is near the point of manufacturing scale-up, and even then it is used not for reaching "go—no go" decisions, but for optimizing the allocation of resources in the scale-up effort. Before that point, given the uncertainties associated with new product development, it would be "lunacy in our business to decide that we know exactly what's going to happen to a product once it gets out." A continuing effort is made to develop and apply financial analysis tools which, rather than govern decision making, can help decision makers deal with the high uncertainties in new drug development in ways that are consistent with "the construct of returning the cost of capital."
- George B. Rathmann at Amgen made a similar point. "You cannot really run the numbers, do net present value calculations, because the uncertainties are really gigantic. . . . You decide on a project you want to run, and then you run the numbers [as a reality check on your assumptions]. . . . Success in . . . a business like this is much more dependent on tracking rather than on predicting, much more dependent on seeing results over time, tracking and adjusting and readjusting, much more dynamic, much more flexible."
- Schoemaker at Centocor works "in zones of gray." He describes his company as being more strategy and event oriented than financial. Even late in the development of a new drug, highly uncertain critical events (e.g., performance of competitive products, results of clinical trials) are such large "swing factors" that future returns simply cannot be reliably projected. Under these circumstances, DCF analysis plays more of a support role than a determinative one, a view echoed at Merck & Co. and Amgen.
- At Analog Devices and Intel, similar views were expressed. Stata made the point that "particularly in a high-tech business, we do not see how you can play the projecting the numbers game." And Moore, at Intel, argued that "you cannot run a very good model with a situation where prices can fall 90 percent in 9 months. . . . You may run the numbers on the model, but intuitively you worst case it and decide that some things are just too darn risky no matter how the numbers come out. . . . In the businesses

where we have to stay in . . . almost all our investments are made because they are strategically important."

- Referring to the $50-million investment that his company is now making to build a new steel processing plant, McConnell at Worthington Industries said that "from an accounting point of view, our Porter plant is not a good idea. An accountant would say it's not a good idea, from a straight dollar payback perspective. But there are so many intangible benefits that it more than pays for itself."

Likewise, there was considerably less emphasis in this group on interest rates as a critical policy issue. Only Merck & Co. cited economic policy as its primary policy concern. In Spiegel's view, the number one issue is "the common enemy . . . the enormous and increasing U.S. national debt." For the three other pharmaceutical companies, policies influencing the "tail end" of the innovation cycle—policies affecting the reimbursement of medical costs and the pricing of prescription drugs, protection of intellectual property, and delays in the issuance of patents and Food and Drug Administration approvals—are of greatest concern. Merck & Co. also believes these are important issues, and its emphasis on the national debt stems in large measure from a concern about the impact of that debt, and efforts to reduce it, on the nation's health policy and health care system.

As for the semiconductor firms in this second group, both Intel and Analog Devices cited R&D policy as critical. Moore (Intel) also emphasized trade policy, while Stata (Analog Devices) argued that there was much more that the federal government could do to promote learning among firms. For example, he cited the problem of quality, and how in the absence of a mechanism for learning across firms, many firms are forced to work through the same problems and mistakes. Nucor and Worthington Industries, meanwhile, mentioned none of these issues and instead suggested that for the steel industry, the costs of environmental regulation—or more precisely, the pursuit of "reasonable" requirements—was most important.

Group 3: The Mixed View

Hurco Companies and Chips & Technologies took an intermediate view on the impact of cost of capital on their businesses.

- Hurco Companies: Brian D. McLaughlin, CEO, argued that for a small machine tool company like Hurco Companies, which faces competition from the likes of the Fanuc-GE joint venture, Mitsubishi, and Siemens, the key to remaining competitive is to avoid competing on a capital-intensive basis (i.e., on the manufacturing of mechanical and microelectronic hardware) and to focus instead on staying ahead in controls, user interfaces, and application software. Hurco Companies thinks of the machine tool as "a shipping crate" for its software and controls. This suggests that cost of capital is not a driving force in this business. On the other hand, the very logic of avoiding the capital-intensive aspects of the machine tool business suggests a fundamental concern about availability and cost of capital. And in two other respects, McLaughlin was quite explicit about the impact of the cost of capital on his business. First, while Hurco Companies is financially sound today, it was in deep financial difficulties in the mid-1980s. Then, the cost of capital was a dominant concern. Second, to the extent that a high cost of capital dampens the climate for investment in manufacturing plant and equipment in this country, it has a direct impact on Hurco Companies. This is why McLaughlin's priority for federal policy is to strengthen the overall economic environment, beginning with a lowering of interest rates. His second policy priority is to strengthen, in part through technology policies, the manufacturing base of the country.

- Chips & Technologies: Gordon A. Campbell, chairman, president, and CEO, reported that while the cost of capital is a concern, it is not one of the top factors constraining his business. "We know the type of products we need; we know the time frame in which we need to develop them; . . . we have a roadmap. I can't blame that [failure to realize those plans] on the cost of capital." On the other hand, the reason the cost of capital is not a critical issue for Chips & Technologies is that the firm very deliberately set out to avoid capital intensiveness. When Chips & Technologies was starting up, the only way to raise enough capital to build its own fabrication facilities would have been "to give away 80 percent of the company." So the general approach became to "do anything you can to avoid becoming capital intensive." Just as Hurco Companies views the machine tool as a "shipping crate," Chips & Technologies views semiconductor fabrication capability as "a commodity." Its devices are now fabricated at about a dozen foundries worldwide. From a policy perspective, Campbell was much more concerned about the general problem of a lack of a cohesive policy regarding the competitiveness of U.S. industry than specifically about interest rates or policies affecting capital formation.

Table A.1 summarizes the responses of the three groups of executives to the first question in the interview: Is the cost of capital an important source of competitive disadvantage?

Question 2: How Do the Firms that Emphasize the Competitive Impact of the Cost of Capital Manage the Problem?

The interviews show that U.S. companies are employing an array of tactics for managing the conflict between the constraining effects of the

TABLE A.1 Summary of Responses

	Cost of Capital an Important Competitive Factor?	Emphasis on DCF in Decision Making for Specific Investments?	Policy?
Group 1			
Texas Instruments	Yes	Yes; less than in past	Capital formation
Armco	Yes	Yes; basis for screening	Trade, environmental, interest rates
Inland	Yes	Yes; basis for screening	Interest rates
Cincinnati Milacron	Yes	Not discussed	Competitiveness
Cross & Trecker	Yes	Not discussed	Export policies
Group 2			
Analog Devices	No	No; uncertainty, strategic	R&D policy, learning
Intel	No	No; uncertainty, strategic	R&D, trade policy
Merck	No	No; uncertainty, strategic	National debt, health policy
Amgen	No	No; uncertainty, strategic	Health policy, patents
Centocor	No	No; uncertainty, strategic	Health policy, patents
Nucor	No	No; strategic factors	Environmental
Worthington	No	No; strategic factors	Environmental
Group 3			
Hurco	Indirectly	Not discussed	Economy, manufacturing
Chips & Technologies	Indirectly	Yes for incremental steps, no for more radical steps	Competiveness

cost of capital and the strategic necessity for continuing investments. The behavior of these companies suggests that the relative cost of capital is not a fixed constraint but a condition that is at least partially manageable.

Texas Instruments, as it struggles to regain leadership in the capital intensive DRAM business, has taken a variety of steps that in Junkins' words, "simulate access to a low cost of capital environment." For example, TI has entered into a series of joint ventures that have in effect made an additional $1 billion available to the company over the next several years.

• In an agreement with the Italian government, TI will invest $500 million over four years to expand its Italian operations. The expansion will include a 4-megabit DRAM manufacturing facility, a new applications research center, and upgrades of existing facilities. The Italian government will contribute nearly $700 million in cash, tax incentives, low-interest loans, and infrastructure improvements (see, for example, *Wall Street Journal*, November 8, 1989, p. A4).

• In an agreement with Acer of Taiwan, Acer will contribute 75 percent of the capital required to build and operate a $250-million DRAM facility. TI will contribute 25 percent plus technology, and will receive all the output of the facility. Acer has the option to purchase up to 50 percent of the output.

• In an agreement with Kobe Steel, Kobe will contribute most of the $350 million required to build a facility that will produce logic and application-specific devices. TI will sell all the output of the plant under its name, even though it is the junior partner in the joint venture.

Other steps taken by TI to lower its cost of capital include sale-leasebacks and conversion of preferred stock; aggressive prosecution of patent violators, which has generated a significant new stream of income estimated to exceed $200 million per year; and efforts to develop a modular approach to semiconductor manufacturing, which if successful, would reduce the required "capital investment by an order of magnitude." At the same time, TI has been investing aggressively in new capacity. In 1990 alone, during what has proved to be a significant downturn in the semiconductor business, it has continued to accelerate its pace of investment, which has been "tearing heck out of the bottom line." But given TI's commitment to remain competitive in the DRAM business, Junkins does not believe TI can afford to cut back on investment during the downturn. Investing through the downturn is thus another way in which TI is responding to the strategic requirement to invest in an environment of high capital costs.

In the steel industry, Armco and Inland Steel Industries have engaged in a not dissimilar set of tactics. Perhaps the most dramatic example is a series of agreements between Inland and Nippon Steel, agreements that have grown into a $1-billion joint venture to build and operate two world-class steel-making facilities—one that manufacturers cold-rolled sheet-steel products and the other that produces coated steel products. The first of the two facilities, which is now in operation, is achieving "unprecedented" productivity and quality. Inland owns 60 percent of the joint venture. As calibration of the magnitude of the effort, the first of the two plants accounts for roughly 20 percent of Inland's total steel production. But what is most significant about the joint venture is how it was financed. It is not carried on Inland's books; the financing for the joint venture is backed solely by the assets of the joint venture itself. Twenty percent of Inland's investment is equity based; the remaining 80 percent is funded through debt by a Japanese institution. Luerssen reports that the interest rate is lower than the rates available from U.S. banks, although higher than the rate available in Japan. More significant than the rate is the capital structure agreed to by the Japanese bank—20 percent equity, 80 percent debt—and the non-recourse financing. The "capital structure of the joint venture looks like that of a Japanese firm." Otherwise, "we never could have gone ahead with the deal."

Armco has taken similar measures. It has entered into several joint ventures, the most dramatic being a joint venture with Kawasaki Steel Company, into which Armco has spun off its carbon steel facilities, which represent about 40 percent of its total capacity. Kawasaki in turn, is investing $500 million in the joint venture, and as 50 percent partner, is assuming half the obligations associated with the spun off facilities as well. Armco projects a need for $1.5 billion in investment in the joint venture, which will be raised by the joint venture itself, not Armco. Like the Inland-Nippon joint venture, the Armco-Kawasaki joint venture is off Armco's books and has access to Japanese financing, which translates to somewhat lower rates but more importantly to a very different and more highly leveraged capital structure.

Apart from these measures for gaining access to foreign capital, the most important other tactic for managing the cost of capital employed in our sample was exhibited by the two smaller firms in Group 3. Both Hurco Companies and Chips & Technologies have devised strategies for competing in their industries without heavy capital investment. Neither manufactures hardware or electronic components for their hardware, thereby greatly reducing their vulnerability to the cost of capital. At Chips & Technologies, "our whole strategy is to avoid capital intensiveness."

Question 3: What Accounts for the Differences in Views on the Cost of Capital?

How do we account for the differences in the executives' views on the competitive impact of the cost of capital? Why do Group 1 executives see it as an important source of competitive disadvantage, while Group 2 executives emphasize other variables—such as quality, ability to stay ahead in process technology, ability to develop new drugs—as weighing more heavily on their ability to compete than possible differences in the cost of capital?

Notice first of all that both views are compatible with economic theory. Group 1 executives argue that they face a competitive disadvantage because given what they see as a relatively high cost of capital, they have difficulty financially justifying investments that their competitors would not have difficulty justifying. But this in no way implies that Group 2 executives are any less concerned about achieving financially justified rates of return than Group 1 executives. They are no less aware of the relationship between the

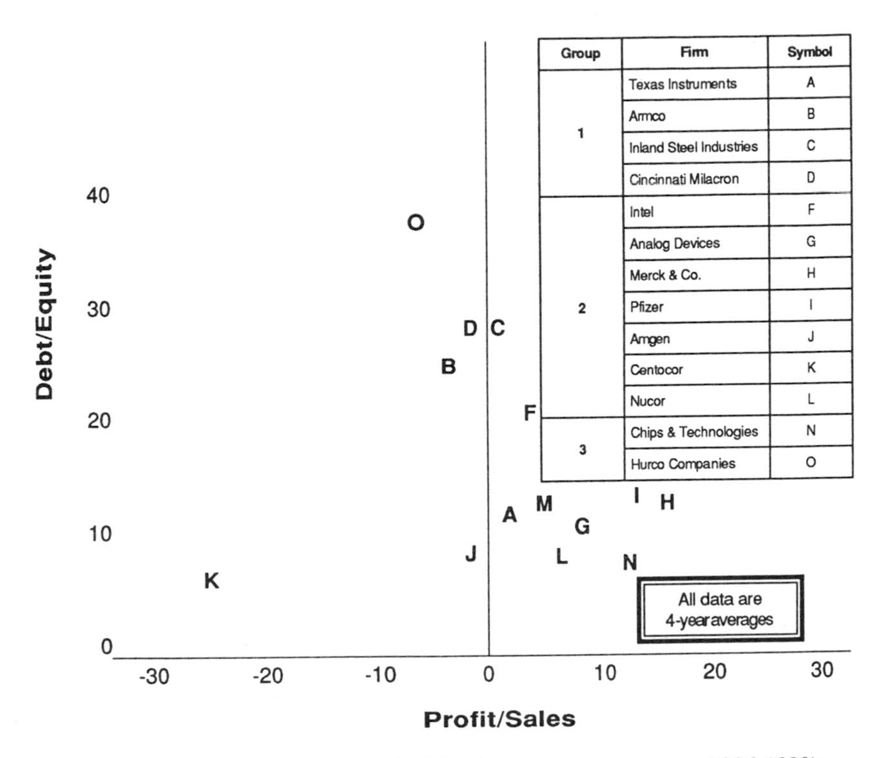

FIGURE A.1 Profitability versus indebtedness (4-year average, 1986-1989).

cost of capital and the economic definition of acceptable returns. They are as likely to calculate the rate of return of potential investments, even if they are not as driven by the results of such calculations in their decision making about specific investments. And if actual returns are any measure of intent, the firms in Group 2 have tended over time to be more profitable than the firms in Group 1. The Group 2 firms in the steel and semiconductor industries have been more profitable than their Group 1 counterparts (see Figure A.1).

In short, it is not plausible to argue that the difference in views about the competitive impact of the cost of capital is due to differences in rationality or concern about shareholder interests. How then do we account for the difference? Why do some executives view the relative cost of capital as fundamental to their ability to compete, and others not?

Industry Type

One possible explanation is that the cost of capital varies from industry to industry, and that in some industries, the difference between U.S. and

TABLE A.2 Responses by Industry

	Group 1	Group 2	Group 3
Machine Tools	Cincinnati Milacron Cross & Trecker		
Steel	Armco Inland		Nucor Worthington
Semiconductors	Texas Instruments	Analog Devices Intel	Chips & Technolgies
Pharmaceuticals		Amgen Centocor Merck Pfizer	

Japanese costs of capital are substantially greater than in others. For example, Merck & Co. (Group 2) doubts that there is a persistent difference in the cost of capital between the United States and Japan, whereas TI (Group 1) estimates that for its industry, the U.S. cost of capital is 75 percent higher than Japan's. Perhaps the different responses simply reflect inter-industry differences in relative cost of capital. This explanation is appealing in its simplicity, but unfortunately, it is only partially supported by the interviews (see Table A.2). All the executives from the pharmaceutical industry believe that differences in the cost of capital are not a critical issue, whereas all the executives from the machine tool industry take the reverse point of view (although Hurco Companies' point of view is a bit more complicated). On the other hand, within the semiconductor and steel industries, different executives expressed different views. This suggests that even if there are inter-industry differences in the relative cost of capital, these differences do not fully account for the differences in opinion about the competitive importance of the relative cost of capital.

Size

Originally, we expected that company size would be related to views about the competitive significance of the cost of capital, since small, growing firms often face the need for more capital than their limited cash flow can provide. But at least in this sample, if there is any relationship between size and views on the cost of capital, it is exactly the opposite of what we had expected (see Table A.3). There are no small firms in Group 1. However, both Group 3 firms—that is, firms that took an intermediate position

TABLE A.3 Responses by Firm Size

	Group 1	Group 2	Group 3
Large:	Texas Instruments Armco Inland Cincinnati Milacron Cross & Trecker	Intel Merck Pfizer	
Small:		Analog Devices Nucor Worthington Amgen Centocor	Chips & Technologies Hurco

on this question—arc small, and while Cincinnati Milacron and Cross & Trecker are large relative to U.S. machine tool makers, they are small in comparison with the other large firms in this sample. If we consider them small firms, the pattern begins to seem unrelated to size. Both large firms and small firms express both points of view with respect to the competitive effects of the cost of capital.

The responses of the three relatively young companies in our survey—Amgen, Centocor, and Chips & Technologies—are especially interesting. On the one hand, Chips & Technologies has built its entire business strategy around its desire to avoid the constraints imposed by the cost of capital. On the other hand, for the two biotechnology firms, the cost of capital was not nearly as important as its availability, as is reflected in their willingness to enter into RDLPs for which the forecast return on investment was 40 percent.

Capital Intensity

Another possible explanation for the differences might be that the firms that emphasize the competitive impact of cost of capital might simply face greater capital requirements. They might be competing in more capital-intensive industry segments or with more capital-intensive strategies than their Group 2 counterparts. We define capital intensity broadly here and include both capital investment and R&D expenditures. In Figure A.2 we have attempted to control for size by dividing capital investment and R&D expenditures by sales. The data again confound our expectations and suggest that capital intensity is not particularly related to executive views on the competitive importance of the cost of capital.

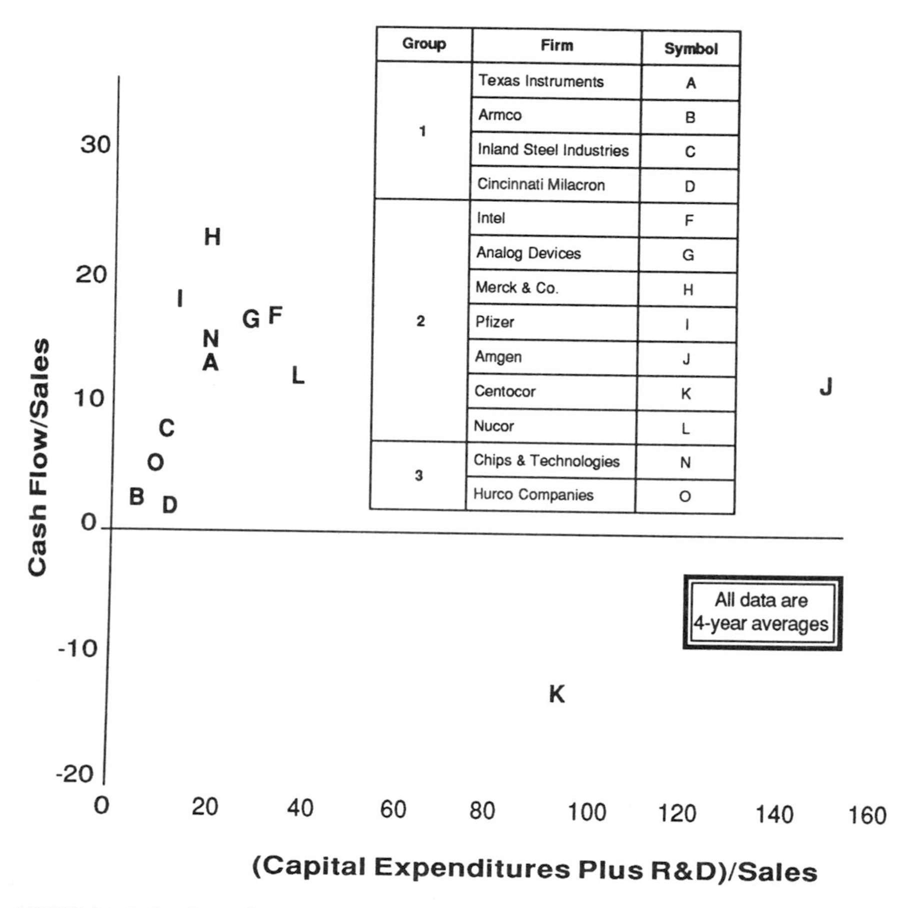

FIGURE A.2 Cash flow versus capital intensiveness (4-year average, 1986-1989).

- In semiconductors, both Analog Devices and Intel (Group 2) have been consistently more capital intensive than Texas Instruments (Group 1) over the past decade.
- In steel, Nucor (Group 2) has been much more capital intensive than either Inland Steel Industries or Armco (Group 1), except during the industry recession in the early 1980s, when levels of spending were roughly comparable. On the other hand, Worthington Industries (Group 2), which is more of a steel processor than a steel maker, is less capital intensive than Inland and Armco.
- The pharmaceutical companies (Group 2) have been more capital intensive than the machine tool companies (Group 1 and 3).

Financial Health and Capital Availability

It might be argued that the real issue is not capital intensity per se, but the financial capacity to satisfy capital requirements. Here, with the very notable exception of the two biotechnology companies in the sample, a pattern does seem to emerge. As can be seen from Figures A.1 and A.2, the firms in Group 1 tend to have lower profitability and cash flow, and higher debt, than the firms in Group 2. Whether we control for industry or not, the firms that view the cost of capital as an important source of competitive disadvantage are in weaker financial condition than those that do not.

The most straightforward interpretation of this result is that financially healthy firms, whatever their level of capital intensiveness, are better able to finance their investment requirements internally (i.e., out of cash flow), and as a result, are less concerned about the costs associated with raising new funds. This is not to say that there is no cost associated with internally generated funds. Reinvested profit is capital raised from shareholders; there is a cost associated with its use, just as there is a cost associated with the use of externally generated capital. But when firms in a financially weaker condition attempt to raise capital externally—and in the case of such capital-intensive industries as steel and semiconductors, raise relatively large amounts of capital externally—they are likely to be confronted with the need to pay increasing costs for that capital. This may help to explain why they express a greater concern about their relative cost of capital. Interestingly, for several of the Group 1 firms, the problem of cost of capital begins to look like a problem of availability of capital. In theory, capital is always available if one is willing to pay a high enough cost, but in practice, firms like Inland, Armco, and TI may find it impossible to raise by conventional means the capital they believe they need to remain competitive. This appears to be why Inland and Armco, for example, have joined with Japanese firms to build their next generation of plants, and have turned to Japanese sources to raise the capital needed to finance their portion of the joint ventures! In theory, capital is always available for a price, but in practice, in these instances, the capital needed to remain competitive simply would not have otherwise been available.

Market Leadership

Closely related to financial health is market leadership. The Group 2 companies tend to be leaders in their respective markets, whereas the market position of the Group 1 companies tends to be problematic.

- In Pharmaceuticals, Merck & Co. is generally considered to be the most successful firm of the 1980s. Amgen and Centocor are among the handful of survivors of the more than 1,000 start-up companies that originally entered the biotechnology field, and after a decade of product development, they appear poised for rapid and profitable growth. And while Pfizer has had some recent difficulties, it is generally believed to have a well-stocked pipeline of new product possibilities.

- In Semiconductors, Intel (Group 2) is the world leader in microprocessor technology, and Analog Devices (Group 2) is widely acknowledged as such in analog signal processing devices. In contrast, TI (Group 1) lost its leadership position in memory devices in the mid-1980s, and is now doing everything it can to overcome the formidable obstacles to regaining leadership in an industry dominated by economies of scale and experience.
- In Steel, the two smaller companies are leaders in their respective market segments. Nucor is generally viewed as the most successful of the "mini-mill" steel companies that have been competing in the lower-value segments of the steel market (e.g., bar and wire rod products) and that increasingly, are posing a threat to the higher-end segments. Worthington Industries is similarly perceived in the steel-processing (as opposed to steel making) segments that it serves. In contrast, our two Group 1 steel makers, Inland and Armco, while they are among the strongest of the domestic integrated steel makers, face formidable foreign competition.
- In Machine Tools, our two Group 1 companies lost their market leadership over the course of the past decade, except in the area of plastic injection molding equipment, where Cincinnati Milacron is acknowledged to be the world leader.

While there seems to be an association between leadership and views on the competitive impact of the cost of capital, how exactly is this relationship to be explained? There appear to be a number of plausible lines of argument. The first is almost tautological. The firm that leads does not have to catch up. Catching up, particularly in businesses where learning curves and economies of scale are as important as they are in capital-intensive industries, becomes prohibitively expensive. The firm must come from behind in a field where the leader is always gaining. It must make the investments required to close the gap while at the same time matching the investments and advances that the leaders are making to enhance their position. The problem becomes all the more difficult if, as is usually the case, the firm striving to catch up is in a weaker financial condition than the firm or firms in the lead. The gap between available capital and required capital becomes more severe, which in turn places all the more importance on the costs and difficulties associated with raising the capital required to close that gap. This is exactly the pattern that seems to have afflicted the steel, semiconductor, and machine tool companies in Group 1. Iverson, chairman and CEO of Nucor (Group 2), described this pattern explicitly as he explained his investment philosophy:

> In steel, you don't have any choice. You have to keep investing continuously. If you don't, you fall behind. And if you fall behind, it is very difficult and extremely expensive to catch up. This was the problem that the integrated steel makers ran into. You wind up highly leveraged, which gets you into serious trouble during downturns.

Another way to look at this association between leadership and lack of emphasis on the cost of capital as a competitive factor is to consider how the companies in Group 2 have preserved or built their strong market positions. All have done so largely on the basis of technology leadership. Intel and Analog Devices are technology leaders in microprocessors and analog signal-processing devices, respectively; Nucor has long been a pioneer in the steel industry and recently invested $270 million in building the world's first continuous thin-slab cast flat-rolled steel-making facility, while Worthington Industries is the U.S. leader in the application of advanced steel processing technology developed in Europe; Merck & Co. is widely considered the most innovative company in the pharmaceutical industry, Amgen and Centocor are among the handful of successful biotechnology companies, and Pfizer appears to be well positioned technologically.

Moreover, if we were to ask the Group 2 executives to name the most important determinant of their ability to maintain their positions of advantage, their answer would be a continuing capacity to innovate. For the pharmaceutical firms, this means continued leadership in development of major new drugs. For the semiconductor firms, it means continued product leadership in microprocessors for Intel and in analog devices for the company of that name. For Nucor, it means advanced steel-making capacity, in addition to low costs and debt; and for Worthington Industries, customer satisfaction, which requires the most advanced steel-processing facilities in the markets they serve. Whether or not these firms are able to maintain in the future the positions they have built in the past will depend on whether they can continue to outinnovate their competitors. And because innovation, and the ability to continue to innovate, are fundamentally important to the well-being of these companies, the secondary importance given to the cost of capital as a competitive factor becomes understandable. The ability of these firms to maintain their technology leadership will have a vastly greater impact on their long-term profitability than any relative differences in the cost of capital. Conversely, if these firms fail in their quest to preserve technology leadership, the impact on their profitability will far exceed the impact of any relative differences in cost of capital. This is why Laubach, president of Pfizer, emphasized that especially with the advent of the generic drugs, the only way to succeed in the pharmaceutical industry—that is, the only way to continue to generate attractive rates of return over the long haul—is to develop new products, and this requires that the firm be willing and able to invest on average, $230 million of R&D for each new drug, and simultaneously to invest in a portfolio of such new drug possibilities. "If we are going to stay in this industry, we've got to play [the new product development game]. If we don't play, we're out. If we play and lose, we are also out, but it is better to play and try to win, than to walk away."

Our Group 1 companies also explicitly begin with the premise that they are committed to their respective businesses. The difference is that they are working from a position of, at best, technological parity. And when firms in a competitive, capital-intensive arena possess roughly comparable technology, the ability to invest in that technology—that is, the magnitude of available resources—becomes a primary determinant of competitive success. The firms compete not on the basis of their innovativeness, but on their ability to keep up with the industry leaders' pace of investment in state-of-the-art (as opposed to leadership) plant and equipment. Under these circumstances, the cost—and we would argue, availability—of capital becomes a dominant concern. Other things being equal, particularly technology, the firm with access to more or cheaper capital, or both, has the ability to outinvest its competitors. Not only do the firms operating under a less attractive capital environment fall behind, but the gap between them and their competitors can grow very quickly.

CONCLUSION

A growing body of evidence suggests that the differences in the time horizons of U.S. and Japanese firms are deeply rooted in the industrial structure and cultures of the two countries. Whether or not there are persistent, real differences in the cost of capital, Japanese firms will continue to behave for some time to come as if they enjoyed a lower cost of capital. One of the inescapable features of global competition in the 1990s and early twenty-first century will be the presence of foreign competitors with long time horizons—competitors who from a U.S. perspective, behave as if they had lower costs of capital.

The ability to compete against firms that behave as if they had lower costs of capital thus becomes a prerequisite for U.S. firms participating in global markets. How executives in capital-intensive marketsk view the competitive effects of the U.S. cost of capital, and why some are less concerned about it than others, become especially interesting questions. The simplest and most general conclusion from our survey is that the way to remain competitive against firms that behave as if they had a lower cost of capital is to stay ahead! This seems simplistic, yet the firms in our survey that do not perceive relative cost of capital as a critical competitive factor are those that have succeeded in preserving leadership in their respective markets. Once leadership is lost in a particular market, the firm that is able to behave as if it has a lower cost of capital—whether or not it actually does—has an obvious advantage. It will be willing to invest at a more rapid clip than its competitors. In other words, to the extent that the industrial and financial structure of Japan enables Japanese firms to behave as if they had a lower cost of capital—the playing field in capital-intensive markets is uneven. Any firms entering that field without some compensating "unfair advantage," without some clear basis for leadership in the marketplace, will inevitably face a losing battle. And once they fall behind, the slope of the field makes catching up all the more difficult.

How have the Group 2 firms in our survey stayed ahead? Apparently by building and then preserving technological leadership. Being state of the art in capital-intensive industries, having technology that is as good as the competition's, does not appear to be good enough, at least on the basis of this limited survey. Technological parity in a capital-intensive business implies that firms are competing on the basis of the rate at which they are willing or able to invest. As long as their structure and culture enable Japanese firms to behave as if they have lower costs of capital, it is difficult to see how any U.S. firm can compete successfully over the long run on these terms. On such an uneven playing field, all that is left as a basis for competition once technology leadership is lost is access to domestic markets, and it is not surprising in these circumstances to see a growth in the number of joint ventures between Japanese firms with capital and U.S. firms with access to domestic markets.

This does not mean that we should neglect efforts to reduce the disparity in time horizons between the United States and Japan. On the contrary, every effort must be made to do so. A broad range of fundamental policy steps will be required. And if the structural and systemic differences that enable Japanese firms to pursue relatively long time horizons are likely to persist, then the important question for the near term becomes; what if anything can firms and the federal government do to offset their effects? This analysis suggests that the answer lies with actions of individual firms and the federal government that enhance or preserve U.S. technology leadership.

NOTES

1. For a recent review of comparative rates of investment and productivity, see Norsworthy, 1990. Some authors now emphasize that the difference in rates of investment applies both to fixed or tangible investments, as well as to invisible or intangible investments. For example, see Hatsopoulos et al., 1988.
2. See, for example, Norsworthy, 1990. An analysis by Dr. B. Catto, Chief Economist at Texas Instruments well illustrates the point. It shows that although cost of debt and cost of equity are roughly comparable for U.S. and Japanese semiconductor manufacturers, the overall cost of capital for the U.S. manufacturers is roughly 75 percent higher because the capital structures of their Japanese counterparts are much more highly leveraged (and the cost of debt is lower than the cost of equity). The average debt/equity ratio for Japanese

semiconductor manufacturers is more than an order of magnitude higher than for U.S. semiconductor manufacturers.

3. The study was designed to maximize the number of interesting contrasts we could make, since most useful information concerning the cost of capital is relative. Moreover, pairs of firms were selected for each category to provide an internal measure of repeatability or consistency. And financial data covering two business cycles were collected for each firm so that the influence of the business cycle could be investigated.
4. For a general description of this approach to raising capital, see Henriques, 1991.
5. See DiMasi et al., 1991.

REFERENCES

Abuaf, N., and K. Carmody. 1990. The Cost of Capital in Japan and the United States: A Tale of Two Markets. Salomon Brothers Financial Strategy Group (July).

DiMasi, J. A., R. W. Hansen, H. G. Grabowski, and L. Lasagna. 1991. The cost of innovation in the pharmaceutical industry. The Journal of Health Economics, Vol. 10 (July) pp. 107-142.

Hatsopoulos, G. N., and S. H. Brooks. 1986. The Gap in the Cost of Capital: Causes, Effects, Remedies. R. Landau and S. W. Jorgenson, eds. Technology and Economic Policy. Cambridge, Mass.: Ballinger.

Hatsopoulos, G. N., P. R. Krugman, and L. H. Summers. 1988. U.S. competitiveness: beyond the trade deficit. Science (July 15):241.

Henriques, D. B. 1991. Disguising the risks of research. New York Times, Business Section (February 3): 14.

Kester, W. C. 1991. Japanese Takeovers. Boston, Mass.: Harvard Business School Press.

Kester, W. C., and T. A. Luehrman. 1989. Real interest rates and the cost of capital. Japan and the World Economy 1.

Kester, W. C., and T. A. Luehrman. 1990a. Cross Country Differences in the Cost of Capital: A Survey and Evaluation of Recent Empirical Studies. Draft (December 12).

Kester, W. C., and T. A. Luehrman. 1990b. The price of risk in the United States and Japan. Harvard Business School Working Paper, 90-050.

McCauley, R. N., and S. A. Zimmer. 1989. Explaining international differences in the cost of capital. Federal Reserve Bank of New York Quarterly Review (Summer).

Norsworthy, G. N. 1990. Why U.S. manufacturers are at a competitive disadvantage. MAPI Policy Review (November 1990) PR-114. Washington D.C.: Manufacturers' Alliance for Productivity and Innovation.

Zielinski, R., and N. Holloway. 1991. Unequal Equities. Kodansha International. Tokyo.

APPENDIX B

Biographical Information on Committee Members

DONALD N. FREY is a professor in the Department of Industrial Engineering and Management Sciences in the Robert R. McCormick School of Engineering and Applied Science at Northwestern University. He has been active in both the academic and corporate workplace during his career, most recently serving as chairman of the board and chief executive officer of Bell and Howell Company from 1971 to 1988. During that time, he also served as lecturer and adjunct professor at Northwestern University and the University of Chicago. Prior to his position at Bell and Howell, Dr. Frey began his career in the metallurgical department in Ford Motor Company's Scientific Laboratory in 1951, and progressed to the position of vice president of the product development group at Ford in 1967. In November 1990 he received the National Medal for Technology award from the President of the United States. Dr. Frey received his B.S. in metallurgical engineering, M.S. in engineering, and Ph.D. in metallurgical engineering from the University of Michigan.

ROBERT C. FORNEY is a retired executive vice president, member of the board of directors, and member of the executive committee of E. I. du Pont de Nemours & Company. Dr. Forney held positions of increasing responsibility in Du Pont, including product manager, director of the Products Marketing Division, general director of the Marketing Division, and vice president and general manager of the Textile Fibers Department. He is a former member of the Board of Governors of the Purdue Foundation and serves as a director on several boards. Dr. Forney received his B.S. and Ph.D. in chemical engineering from Purdue University.

MARTIN GOLAND is the president of Southwest Research Institute. He is active in numerous scientific advisory groups at the national level and has broad experience in aircraft design, applied mechanics, and operations research. Previously, Mr. Goland held positions in Midwest Research Institute and the Curtiss-Wright Corporation and was an instructor at Cornell University. He is the author of more than 60 papers on structures, aerodynamics, dynamics, mathematics, engineering analysis, research administration, and other subjects. He was awarded the prestigious Hoover Medal in 1987 and the U.S. Army's Outstanding Civilian Service Award in 1988. Mr. Goland received his B.S. in mechanical engineering from Cornell University.

GEORGE N. HATSOPOULOS is the founder, chairman of the board, and president of Thermo Electron Corporation. The comany's principal businesses include manufacturing of environmental and analytical instruments, alternative-energy power plants and prepackaged cogeneration systems, industrial process and power equipment, and biomedical products. Dr. Hatsopoulos served on the faculty of MIT from 1956 to 1962 and continued his association with the Institute, serving as Senior Lecturer until 1990. He is a member of the governing Council of the National Academy of Engineering, and a vice chairman of the American Business Conference, a member of the Executive Committee of the National Bureau of Economic Research. He served on the Board of the Federal Reserve Bank of Boston from 1982 through 1989 and as its chairman in 1988 and 1989. He has testified at numerous Senate and congressional hearings on national energy policy and capital formation, and he has served on many national committees on energy conservation, environmental protection, and international exchange. Dr. Hatsopoulos received his education at the National Technical University of Athens and at the Massachusetts Institute of Technology, where he received his B.S., M.S., M.E., and Sc.D. degrees in mechanical engineering.

TREVOR O. JONES is chairman of the board of Libbey-Owens-Ford Company and also president of the International Development Corporation of Cleveland, Ohio. A native of Maidstone, England, Mr. Jones started his U.S. engineering career with General Motors in 1959, where he spent 19 years working in aerospace activities and in 1970 was charged with bringing aerospace technology to automotive safety and electronic systems. He became director of GM's newly organized Automotive Electronic Control Systems group in 1970, was appointed director of Advance Product Engineering in 1972, and became director of GM's Proving Grounds in 1974. Mr. Jones was employed by TRW in a number of executive positions, including vice president of engineering for TRW Automotive Worldwide, group vice president and general manager of TRW's Transportation Electrical and Electronics

Group, and group vice president of Strategic Planning, Business Development, and Marketing for the Automotive Sector. Mr. Jones completed his formal engineering education in the United Kingdom at Aston Technical College and Liverpool Technical College.

HENRY KRESSEL is currently managing director of Warburg, Pincus & Co., a diversified venture capital and financial management firm. From 1959 to 1983 Dr. Kressel occupied various positions of increasing responsibility at the RCA Laboratories, Princeton, New Jersey. In 1979 he became staff vice president for solid-state research and development. His responsibilities encompassed integrated circuits and optoelectronic devices and systems. He led the development of many semiconductor devices, accomplishing a succession of breakthroughs, most notably in the field of transistors and optoelectronics. Notable achievements include pioneering the development and commercial introduction of the heterojunction semiconductor laser technology. A graduate of Yeshiva College in physics, Dr. Kressel received an M.S. degree in applied physics from Harvard University, an M.B.A. from the Wharton School of the University of Pennsylvania, and a Ph.D. in materials sciences from the same university.

JOHN R. MOORE is a retired vice president and general manager of the Electro-Mechanical Division of the Northrop Corporation and is currently a consultant to high-technology industries. Prior to that, he was president of Actron Division and a corporate vice president at McDonnell Douglas Corp. and served on the faculty at both Washington University and the University of California, Los Angeles. In 1966 he was elected executive vice president of North American Aviation and a member of the board of directors and after the North American Aviation—Rockwell Standard merger, he became head of the North American Aviation part of North American Rockwell, a member of the board, and a member of the executive committee. Mr. Moore is a graduate of the G.E. Advanced Course in Engineering and holds a degree from the UCLA Executive Program and a B.S. from Washington University.

JOHN WILLIAM PODUSKA, SR. is the founder, chairman, and chief executive officer of Stardent Computer Inc. Before founding Stardent, Dr. Poduska was the founder and chairman of the board of Apollo Computer, Inc., and a founder of Prime Computer as well as the vice president of research and development. He was the director of the Honeywell Information Science Center, chief of the Man-Computer Systems branch at NASA's Electronic Research Center, and an assistant professor of electrical engineering at the Massachusetts Institute of Technology. Dr. Poduska holds Ph.D., M.S., and B.S. degrees, all from the Massachusetts Institute of Technology.

JAMES BRIAN QUINN is the William and Josephine Buchanan Professor of Management at the Amos Tuck School of Business Administration at Dartmouth College. He joined the Tuck faculty in 1957 and is an authority in the fields of strategic planning, the management of technological change, and entrepreneurial innovations. Professor Quinn has held fellowships from the Sloan Foundation, the Ford Foundation, and the Fulbright Exchange Program. In addition to consulting with leading U.S. and foreign companies and publishing extensively on corporate policy issues, he has the distinction of recently being named the Dean of The International University of Japan. Dr. Quinn earned a B.S. from Yale, an M.B.A. from Harvard, and a Ph.D. from Columbia University.

SHELDON WEINIG is the founder and chairman of Materials Research Corporation (MRC), a wholly owned subsidiary of Sony USA Inc. MRC is a multinational company supplying sophisticated materials and equipment to the electronics and computer industries. Before starting the company, he was on the faculty of Columbia University and New York University. In 1988 the government of France awarded Dr. Weinig the rank of Chevalier dans l'Ordre National de la Légion d'Honneur. Dr. Weinig was a member of the United States-Japan Scientific Exchange Committee and a member of President Reagan's Board of Advisors on Private Sector Initiatives. He received a B.S. in metallurgical engineering from New York University and an M.S. and D.Sc. in metallurgy from Columbia.